SCIENCE
Resource Guide

Edited by
Rozanne Lanczak Williams

Contributing Writers
Kathy Burdick
Marcia Fries
Kim Jordano
Gina Lems-Tardif

Project Directors
Carolea Williams
Rozanne Lanczak Williams

Art Director
Tom Cochrane

Designed by
Moonhee Pak

Illustrated by
Kathleen Dunne

Photographed by
Michael Jarrett

Special, heartfelt thanks to the many teachers and students across the country who have contributed their wonderful ideas and projects for this book.

CTP ©1996, Creative Teaching Press, Inc., Cypress, CA 90630

TABLE OF CONTENTS

ACTIVITIES

LIFE SCIENCE

EARTH SCIENCE

PHYSICAL SCIENCE

INTRODUCTION

The *Learn to Read Resource Guide* provides a wealth of ideas and activities for integrating the *Science* emergent reader series in a balanced literacy program. The *Learn to Read Science* series includes 24 books that have been carefully developed to provide emergent readers with text they can successfully read on their own. The text reflects science concepts commonly taught in primary grades.

- hands-on science activities
- experiments
- interactive science displays
- text innovations
- class books
- individual student books
- story dramatizations
- oral reports

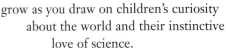

Children learn to read by reading and to write by writing. They develop skills, strategies, and fluency in a language-rich environment where they have many, varied opportunities to read, write, listen, and speak. The activities and ideas in this guide will pro-

vide you with a wide variety of motivational activities and innovative ideas to support the beginning reader's science learning and reading skill development.

The ideas and projects were created by kindergarten, first-, and second-grade teachers and their students all across the country. Their projects are photographed throughout the guide and include:

- puppets
- games
- pocket chart activities
- story murals
- graphing
- collaborative projects

Children who think of themselves as readers and writers, and whose every attempt is encouraged and supported, develop the confidence to take on new reading and writing challenges. By integrating literacy learning with science concepts, you will be able to implement a program customized to your students' needs. Reading skills will naturally grow as you draw on children's curiosity about the world and their instinctive love of science.

RESOURCE GUIDE COMPONENTS

For each of the 24 books in the *Learn to Read Science* series, there are three pages of information and extension activities. Each three-page section includes the following special features:

WRITING FRAMES

In this section, several writing frames modeling patterned language in the book are listed. For more information about using writing frames, see page 8.

SCIENCE CONCEPTS

Science concepts related to the text and illustrations in the book are listed. Concepts include topics most commonly taught in primary grades.

RELATED SKILLS

Opportunities abound for teaching specific skills and reading strategies within the context of the series. Look at this section when planning instruction and addressing children's reading difficulties. For more information about related skills, see pages 6–7.

PHOTOGRAPHS

A picture is worth a thousand words. Bright, colorful photographs of projects created by K–2 students appear through-out this guide and provide extra clarity to written directions.

SYNOPSIS

A short sentence and two-page book spread serve as a reminder of the book's content.

ACTIVITIES

Unique and creative activities to extend science learning, reading, and writing make up this section. Projects include hands-on science activities, experiments, student-made big books, individual student books, murals, pocket chart activities, wall stories, and more.

MATERIALS

Easy-to-find materials are listed for each activity. Items common to all classrooms, such as scissors, crayons, and glue, are listed as art supplies. Collage materials include buttons, confetti, fabric and paper scraps, wiggly eyes, pipe cleaners, stickers, yarn, glitter, sequins, and dried macaroni.

LITERATURE LINKS

This section includes a list of books related to the themes and content of the *Learn to Read* series and the activities presented in this guide.

LEARNING A SKILL

One skill from the Related Skills list is developed for the book. Look at the great ideas in this section for ways to incorporate specific skill instruction.

LINKING SCHOOL TO HOME

These take-home activities provide a non-threatening invitation to parents to become part of the classroom community. They encourage communication between home and school, help children connect home and school learning, and provide lots of opportunities for children to share and reinforce new skills.

ABOUT THE *LEARN TO READ*
SCIENCE SERIES

Lots of reading! Lots of science!

The *Learn to Read Science* series is designed as a flexible resource for your early literacy program. The books have been written and carefully developed to provide emergent readers with text they can successfully read on their own. The engaging stories, along with colorful and appealing illustrations, make reading a fun and enjoyable experience.

The *Learn to Read Science* series consists of 24 student-sized books for emergent readers and 24 matching big books. Their content reflects science concepts commonly taught in primary grades. Themes such as the five senses, living vs. nonliving, plants, animals, and environments are explored in the text and illustrations. The books are written on two levels:

Level I books contain eight pages of easy-to-read text. Usually one line of text appears on each page. There is one language pattern with no more than two changes on each page.

Level II books contain sixteen pages of slightly more difficult text. One or two lines of text appear on each page. The language pattern may change once throughout the story but remains highly repetitive.

The following special features in the *Learn to Read Science* series give maximum support to the beginning reader:

- Repetitive, predictable story lines provide instant success.

- Engaging stories with satisfying endings promote reading for meaning, rather than just sounding out words.

- Colorful illustrations in a variety of artistic styles closely match the text and provide added support.

- Large print is clear and well-placed.

- Natural language patterns (what the reader is used to hearing) move the reader easily through the text.

- Easy and fun activities on the inside back cover extend language and science learning.

BUILDING A BALANCED LITERACY PROGRAM

Focusing on Skills and Strategies

The development of skills and strategies is an ongoing part of a balanced literacy program and occurs within the context of the reading and writing children are doing in the classroom. Skills can be taught formally when children are experiencing specific difficulties or when you antici-pate difficulty with a particular text. Skills are tools learn-ers use to make sense of a story when they read and to communicate effectively when they write. Most impor-tantly, skills become strategies when learners apply them to solve their reading and writing difficulties. Developing strategies should be the focus of all skill instruction.

The components of a balanced literacy program include:

- reading aloud
- shared reading
- guided reading
- independent reading
- writing aloud
- shared writing
- guided writing
- independent writing

Reading

READING ALOUD

Reading aloud is an important part of a balanced literacy program. Read to children several times a day in the class-room, and encourage parents to spend at least fifteen min-utes a day reading to their children at home. Reading aloud makes a significant impact on the developing read-ing skills of young children. It builds comprehension, vocabulary, and listening skills, and exposes children to good literature written on a level higher than their instructional level.

Throughout this guide, Literature Links provide lists of books related to the theme and content of those in the series. Enrich your program by choosing read-aloud titles that extend student learning. Children will gain informa-tion and science knowledge they can access when working on their own. For example, by reading books such as *From Seed to Plant* by Gail Gibbons and *Pumpkin Pumpkin* by Jeanne Titherington, you introduce background knowl-edge and vocabulary relating to the life cycle of plants. This extends learning in the *Learn to Read* books *The Seed Song* and *See How It Grows*.

SHARED READING

Shared reading is a powerful tool for teaching children what reading is all about. Children at all developmental levels are invited to join in the reading of a big book, poem, chant, or pocket chart story. Print is enlarged on shared reading material in order to encourage participation by the whole group. Modeling and child participation occur simultaneously. The emphasis during these sessions is on the joy and satisfaction of reading.

Big books in the *Learn to Read* series are designed primarily for shared reading with emergent readers. Use the repetition, rhyme, and predictable sentence patterns in the text, along with the strong support from illustrations, to lead beginning readers through successful reading experiences. Children enjoy reading the big books again and again during shared reading, and they become favorite choices during independent reading.

Use previously-read big books for specific skill instruction or to teach science concepts. For example, *I See Colors* is a great book to teach color words and the high-frequency words *I* and *see*. To integrate science concepts, discuss the illustrations in terms of observing and classifying matter.

GUIDED READING

During guided reading, work with small groups of children who each have a copy of the same book. A guided reading session is a good time to model and reinforce emergent-level strategies such as one-to-one correspondence, return sweep, locating known and unknown words, letter/sound correspondence (phonics), context clues, and visual searching.

As children develop fluency, give them a book they haven't read before that matches their instructional level. Have each child work through the text while getting your and other readers' support. Children discuss the strategies that help them solve reading problems. This is where the real work of reading occurs. After several successful readings of the book, children can take the book home to read to parents.

INDEPENDENT READING

Emergent readers need many opportunities to read independently. Create a print-rich, reader-friendly classroom by making the following materials accessible:

- big books from previous shared reading sessions
- little books mastered during guided reading
- student-created books modeled after shared big books
- previously introduced pocket chart sets
- wall stories, story murals, and poetry charts
- trade books with text suitable for emergent readers

Writing

Reading and writing are inseparable in a balanced literacy program. They are mutually supportive processes—growing expertise in one area influences the other. Encourage emergent readers to write through writing aloud, and shared, guided, and independent writing sessions.

WRITING ALOUD

Write on a chalkboard or chart in front of children and "think aloud" about the text as you write. This provides a powerful model on how to write, and exposes children to writing conventions such as spacing, punctuation, and spelling. Many teachers write the morning message "aloud" (a brief description of what's happening in the classroom or other noteworthy events).

SHARED WRITING

During a shared writing session, students write with you—it is a collaborative effort. As you guide the process, children supply ideas and input. Children at all developmental levels are invited to participate. Shared writing is a good time to write original stories, poems, class news, information books, or about shared experiences such as guest speakers or field trips. Use shared writing to create innovations and retellings of books children enjoyed during shared reading. For frames relating to specific book titles, refer to Writing Frames sections in this guide.

GUIDED WRITING

During a guided writing session, the child does the writing, while receiving your and other children's support and guidance. This is where the real work of writing occurs. On the emergent level, the guided writing session may be fairly structured. For example, group members could repeat and write the same sentence of a writing frame. You may comment on what the writers are doing correctly and supply missing elements to complete the sentence.

INDEPENDENT WRITING

A language-rich environment is not complete without lots of opportunities for children to write on their own. Encourage writing with journals, reading response logs, dramatic play centers with writing supplies, classroom mailboxes, student writing boxes, and observation journals in the science center. The simple text and patterned language in *Learn to Read* books provide a secure and inviting framework for children's written responses. After they read the books, some children will spontaneously adopt the language pattern and write their own versions.

RECOMMENDED READING FOR TEACHERS

Bialostok, Steve. *Raising Readers*. Peguis, 1992.

Fisher, Bobbi. *Thinking and Learning Together: Curriculum and Community in a Primary Classroom*. Heinemann, 1995.

Raymond, Allen (publisher). *Teaching K–8: Professional Magazine for Teachers*. Early Years, Inc.

Routman, Regie. *Invitations: Changing as Teachers and Learners*. Heinemann, 1991.

WHAT'S GOING ON?

A little girl uses her senses
to guess the special event
that is being planned.

MY FIVE SENSES PUZZLE

Hands-on fun with the five senses

A bright idea and project from Tere
Fieldson and her kindergartners,
Hopkinson School, Los Alamitos, California

Make a simple puzzle sheet and reproduce it on white
construction paper or tagboard. Have each child draw and
color something she or he experiences through the five
senses in the appropriate spaces. Then have children cut
apart the puzzles and place the pieces in labeled envelopes. Children will love working on the puzzles and exchanging them with friends.

Materials
- ✓ teacher-made puzzle sheet (one per child)
- ✓ tagboard or white construction paper
- ✓ art supplies

SOCK BOX

Exploring the sense of touch

Cut a hole as large as your fist in one side
of the box. Tape the open edges of the sock
around the hole. Cut off the toe of the sock. Fill
the box with items of various textures and shapes. Seal the
box with duct tape.

Invite children to put their hands in the sock, pick up an
item inside the box, and use their sense of touch to identify
the item.

Materials
- ✓ large shoe box
- ✓ large sock
- ✓ wide masking or duct tape
- ✓ variety of items with different textures and shapes (pencil, matchbox car, golf ball, tennis ball, cotton ball, ruler, block)

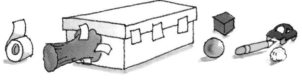

SCIENCE CONCEPTS

- People use their senses to experience the world around them.

- Senses are associated with body parts.

- People use their senses to get information and to draw conclusions.

WRITING FRAMES

I smell a _____.

I taste _____.

I see _____.

I feel a _____.

I hear _____.

RELATED SKILLS

- italics

- drawing conclusions

- picture details

$\mathcal{A}$ctivity DO YOU HEAR WHAT I HEAR?
Exploring the sense of sound

Materials

✓ tape recorder
✓ blank cassette tape
✓ various noise-making items (musical instruments, pan lids, buzzers, bells, alarm clocks, radio)
✓ drawing paper

Collect different noise-making items and record their sounds on a blank tape. Use your voice to make imitations of animals and other sounds. Say, *Sound one,* and record the first sound. Wait five seconds, then record the sound again. Repeat the process as many times as you wish.

After making the tape, give each child six to ten half-sheets of drawing paper. Have children label the first sheet *Sounds I Hear.* Play sound one and ask children to draw a picture of what they hear on another sheet. Repeat for each sound. When finished, invite children to collect their pages and staple them together to make their own books. Words and sentences can be added.

Invite children to experiment with the tape recorder to make their own sound tapes. Children could also use classroom objects and make noises behind a flannel board for partners to guess.

$\mathcal{A}$ctivity MYS-"TASTE"-ERY
Exploring the sense of taste

A bright idea from Betsy McCraine, first grade teacher, Lake Travis School, Austin, Texas

Materials

✓ blender
✓ small paper cups
✓ plastic spoons
✓ variety of foods (fresh bananas, peaches, strawberries; cooked peas, broccoli, potatoes)
✓ water with dissolved sugar, salt, or vanilla
✓ food coloring (optional)

Puree the food items separately in the blender. Place the food and flavored water in separate paper cups for each child. If you wish, you may use food coloring to change the color of the foods to provide an extra taste-testing challenge. Invite children to taste the food and try to guess each one.

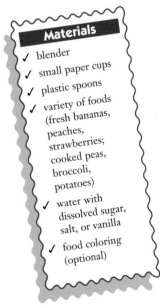

LEARNING A SKILL

Italics

Writers often use italics to call
attention to a specific word. Ask
children to listen carefully as
you read one of the sentences
from the book, first normally and
then emphasizing the italicized
word. Discuss with children how
emphasizing the italicized word changes the sound of the
sentence. Ask why they think the author of *What's Going
On?* used italics in the book. Have children experiment
reading sentences with and without italicized words and
phrases. They can add the same element in their own
writing by underlining words.

Materials
✓ sentence strips or writing paper

Kim Jordano's summer school class at Rossmoor School in Los
Alamitos created an innovation to *What's Going On?*

LINKING SCHOOL TO HOME

Five senses walk

Send home a simple activity sheet
and invite parents and children to
take a short "five senses walk"
around the house or neighborhood. Ask them to record things
they see, hear, taste, touch, and
smell on the activity sheet. Have
children bring the sheet back to
school the next day to share or use as a resource
to write a five senses poem.

Materials
✓ activity sheets
✓ art supplies

LITERATURE LINKS

A, B, See! by Tana Hoban

All the Better to See You With! by Margaret Wild

I Am Eyes Ni Macho by Leila Ward

I Can Tell by Touching by Carolyn B. Otto

My Five Senses by Aliki

The Noisy Book by Margaret Wise Brown

*You Can't Smell a Flower with Your Ear: All About Your Five
Sense*s by Joanna Cole

Your Amazing Senses by Atie Van der Meer

WHERE ARE YOU GOING?

One child leads another on a guessing game using each of the five senses.

SCIENCE CONCEPTS

- People use their senses to experience the world around them.

- Senses are associated with body parts.

- People use more than one sense at a time.

- People use their senses to get information and to draw conclusions.

WRITING FRAMES

I'm going where I can _____.

We went where _____.

I can (see, smell, hear) _____.

RELATED SKILLS

- drawing conclusions

- thought bubbles

- italics

- punctuation: *quotation marks*

- contraction: *I'm*

- sensory description

ACTIVITY

WHERE DID WE GO?
Class book

A bright idea and project from Anne Askay and her kindergartners, Hopkinson School, Los Alamitos, California

Materials
- ✓ construction paper
- ✓ easels
- ✓ painting supplies

This is a great activity to do after a field trip or a walk around the school grounds or neighborhood. After the trip, have children brainstorm a list of what they saw, felt, smelled, touched, and heard. Decide on which phrases everybody likes and write them on sentence strips using the frame: "We went where we felt . . . saw . . . heard . . . ," and so on. Combine them with children's easel paintings for a creative class book.

CAN YOU MATCH THE SOUNDS?
Listening activity

Place each item in two of the ten milk cartons so that two contain rice, two contain beans, and so on. Tape the cartons closed and cover with decorative self-adhesive paper. Have children shake the containers to hear which ones match.

Materials
- ✓ 10 empty half-pint milk cartons
- ✓ rice
- ✓ beans
- ✓ jingle bells
- ✓ wooden beads
- ✓ marbles
- ✓ decorative self-adhesive paper (optional)

TOUCH AND FEEL BOARD
Exploring the sense of touch

Make a Touch and Feel Board by gluing different textured materials to tagboard.

Materials
- ✓ sheet of tagboard (11" x 17" or larger)
- ✓ textured materials (cotton, felt, flannel, corduroy, sandpaper, small stick, button, pebbles, feather)
- ✓ Tacky glue
- ✓ blindfold

Place the board in a learning center set up for pairs of children. Have them take turns wearing a blindfold and guessing each material. Partners can play a guessing game—what feels soft, hard, scratchy, tickly, etc. Children can also try a scavenger hunt. Pick a texture

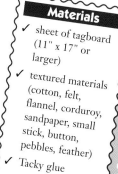

on the board and set a timer for one or two minutes. Partners then try to find something else in the classroom with the same texture in the allotted time.

I HAVE TO HAND IT TO YOU!
Individual student books

Place the listed materials in a writing center. Children can trace and cut out book covers and pages using hand-shaped patterns. Use tagboard for covers and writing paper for inside pages to make a book entitled *I Can Touch*. Each page spread will have the writing frame "I can touch _____" on the left and a piece of one listed material on the right. More advanced writers can add another sentence about how the material feels.

Materials
- ✓ tagboard
- ✓ cotton balls
- ✓ sandpaper
- ✓ flat balsa wood
- ✓ corrugated cardboard
- ✓ velvet ribbon
- ✓ sponge pieces
- ✓ writing paper
- ✓ art supplies
- ✓ hand-shaped patterns

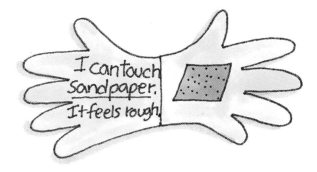

LEARNING A SKILL

Drawing conclusions

Using the frame from the story, describe to children different places you are thinking about. For example, you might say, *I'm thinking of a place where I can feel fuzzy peaches . . . where I can hear music playing . . . where I can smell a ripe melon . . . where I can see lots of food . . . where I can taste a food sample.* From your clues, children should conclude that you're thinking about the supermarket.

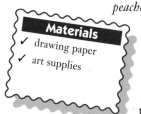

Materials
- ✓ drawing paper
- ✓ art supplies

Have children make their own riddle books. Give each child eight half-sheets of drawing paper, pencils, crayons, and markers. Staple the papers together on the left to make a book. Using the frame, have each child write a riddle about a place he or she remembers visiting. On the last page, have children draw and label the place.

Use the following writing frame:

> *Where Am I?*
> *I can see _____.*   *I can smell _____.*
> *I can hear _____.*   *I can touch _____.*
> *I can taste _____.*   *Where am I?*
>   *At the _____.*

Put the riddle books in a book box decorated with question marks. Children will enjoy guessing each others' riddles by drawing conclusions based on the sentences.

LINKING SCHOOL TO HOME

Riddle books

Have children take home their riddle books to share with family members. Have the whole family make a riddle book about a place they visited together. Invite children to bring the books to school for sharing.

Materials
- ✓ riddle books from *Learning a Skill* activity
- ✓ drawing paper
- ✓ art supplies

LITERATURE LINKS

Fingers and Feelers by Henry Pluckrose

I Spy (series) by Jean Marzollo

I Was Walking Down the Road by Sarah Barchas

I'm Deaf and It's Okay by Lorraine Aseltine

Is It Rough? Is It Smooth? Is It Shiny? by Tana Hoban

Keep Looking! by Millicent Selsam and Joyce Hunt

Listen . . . What Do You Hear? by Jennifer Rye

Look . . . What Do You See? by Nicholas Wood & Jennifer Rye

My Feet by Aliki

Nicki's Walk by Jane Tanner

Touch . . . What Do You Feel? by Nicholas Wood

What Do Animals See, Hear, Smell and Feel? by Victor Waldrop

IS IT ALIVE?

Distinguish between living
and nonliving things in a
fun guessing game.

Is the tree alive?   Yes, it is.

IT'S ALIVE!
Wall story

A bright idea and project from Cori
Giacchino and her kindergartners, Los
Alamitos School, Los Alamitos, California

Place an assortment of living and nonliving objects on a
table. Hold up each object and discuss how we know if it
is living or nonliving. As children discover the characteristics of living things, list them on chart paper.

Materials
✓ living and nonliving
  objects (plant, rock,
  spider, book, fish,
  pencil, apple)
✓ chart paper
✓ butcher paper
✓ easels
✓ painting supplies

Living Things
eat
grow
move
reproduce
drink water

Ask each child to make an easel painting of one living or nonliving thing. Follow the
language pattern in *Is It Alive?* to write text for each page. Children can refer to the
chart and add one characteristic of living things to their page.

SCIENCE CONCEPTS

• All things are living or
  nonliving.

• Living things can be
  classified as plants
  or animals.

• Living things take in
  nutrients, grow, and
  reproduce.

• Nonliving things may
  have characteristics of
  living things.

WRITING FRAMES

Is the _____ alive?

The _____ is alive.

The _____ is not alive.

The _____ is alive, but
the _____ is not alive.

RELATED SKILLS

• question and answer
  format

• sorting and classifying

• punctuation: *question
  marks*

PICTURE THIS
Making posters

A bright idea from Judi Hechtman, first grade teacher, University School, Indiana, Pennsylvania

Ask children to create a detailed picture of themselves and their friends or family doing an activity they enjoy. Encourage them to include lots of detail. Cut labels in half and write "living" or "nonliving" on them. Invite children to identify the things in their pictures as "living" or "nonliving" with the appropriate labels. Bind the pictures into a class book.

Materials
- ✓ 12" x 18" drawing paper
- ✓ file folder labels
- ✓ art supplies

. . . .

LIVING OR NONLIVING?
Making comparisons

A bright idea from Judi Hechtman, first grade teacher, University School, Indiana, Pennsylvania

Divide the class into small groups. Give each group one bowl of guppies, several Gummi fish, and fish food. Have groups observe each type of fish (living and nonliving). Invite them to try to feed both types of fish and make additional observations.

When groups have had ample time to observe, have everyone share their observations. Record children's responses on a Venn diagram. When the diagram is complete, invite children to discuss the characteristics of living and nonliving things.

Materials
- ✓ Gummi fish
- ✓ several small fish bowls containing live guppies
- ✓ fish food
- ✓ large Venn diagram

WHY IS IT ALIVE?

Classification chart

Take a walk with children on the school grounds or in the neighborhood. Have children carry notepads and pencils to record living and nonliving things they see. After returning to the classroom, make a master list on the chalkboard of everyone's observations. Invite each child to pick two or three items from the list to draw on index cards. Use markers and additional index cards to write labels for each item. Make two headers—*Living* and *Nonliving*—to place at the top of the pocket chart. Mix up all the cards and labels and distribute evenly to children. Invite children to take turns placing pictures and labels in the correct categories.

Materials
- ✓ notepads
- ✓ pocket chart
- ✓ index cards
- ✓ art supplies

Extend learning by having children review the pocket chart and brainstorm characteristics of a living thing, such as "it grows," "it moves," "it eats," and "it reproduces." Write these characteristics on index cards. Choose one item from the pocket chart and place all the cards next to it that describe what makes it a living thing.

LEARNING A SKILL

Writing questions

Children can learn to write questions and answers by changing word order in a sentence. Use word cards placed in a pocket chart to make sentences modeled after the text of *Is It Alive?* Show children how changing word order can make sentences a question or an answer. For example, write the question "Is the turtle alive?" Then move the word *is* to make the answer "The turtle is alive." Point out that the punctuation should change. Practice changing more sentences using examples from the book.

Materials
- ✓ pocket chart
- ✓ word cards

LINKING SCHOOL TO HOME

Living and nonliving cards

Send home several index cards with each child, along with simple directions asking help from family members to find pictures of living and nonliving things. Suggest using old magazines. Glue the pictures to the index cards and make accompanying labels. Have children return their cards to school to add to the pocket chart from the *Why Is It Alive?* activity.

Materials
- ✓ index cards
- ✓ magazines
- ✓ art supplies
- ✓ pocket chart

LITERATURE LINKS

Animal, Vegetable, Mineral: Poems About Small Things by Myra Cohn Livingston

Animal, Vegetable, or Mineral? by Tana Hoban

Big Ones, Little Ones by Tana Hoban

Rosata by Holly Keller

The Snail's Spell by Joanne Ryder

Under Your Feet by Joanne Ryder

WHO LIVES HERE?

Where animals live and what they eat are explored in this guessing game book.

ANIMAL MURALS
Small-group activity

Divide children into groups of three or four. Have each group paint a different habitat on butcher paper. Then invite each group member to paint an animal on drawing paper. When the paint is dry, outline animals with black marker. Children can cut them out and attach them to the appropriate background.

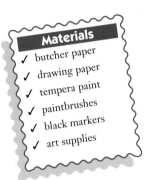

Materials
- ✓ butcher paper
- ✓ drawing paper
- ✓ tempera paint
- ✓ paintbrushes
- ✓ black markers
- ✓ art supplies

Have each child create a small booklet about his or her animal by writing facts or a riddle. Then have children make covers and attach their booklets to the mural with yarn. Classmates will enjoy reading about the animals and locating them on the mural.

Project by first graders at Lee School, Los Alamitos, California

DEAR PET, YOU ARE INVITED . . .
Observing animals

Invite children to bring small, contained pets to school for a few days. Set up stations where children can take turns visiting and observing the animals. Place a science journal at each station where children can draw, dictate, or write about their observations. Observations can include what and how the animals are eating, what their homes look like, how the animals move, and so on. Along with fish, turtles, birds, and frogs, include a container with garden snails, crickets, pill bugs, and other fascinating backyard critters.

Materials
- ✓ small, contained pets and insects
- ✓ science journals

READ THE CLUES
Class pop-up book

A bright idea and project from Kathy Dahlin and her first graders, Cooper School, Superior, Wisconsin

Following the writing frame in *Who Lives Here?*, have children write text for a class pop-up book about animals and their habitats. Children will enjoy reading the clues and guessing the animals.

Directions for pop-up book pages:

1. Fold a sheet of construction paper in half and cut two lines in the center of the fold about 1 1/4" long and 2" apart.

2. Open the page and push the tab inside.

3. On drawing paper, draw, color, and cut out the object that will pop up. Glue the object to the tab, making sure it does not extend below the tab.

4. Draw and color the background and add text.

5. Assemble the book by gluing the completed pages together.

6. Using one large sheet of construction paper or tagboard, make a cover and glue it to the front and back pages of the book.

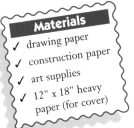

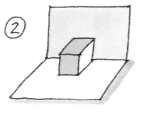

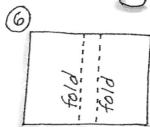

WILD ANIMAL PARK
Three-dimensional display

A bright idea and project from Marcia Fries and her multi-age primary class, Lee School, Los Alamitos, California

Use a black marker to divide and label the cardboard or plywood base into four habitat sections (e.g., jungle, ocean, forest, desert). Divide children into small groups and assign each group a different habitat.

Invite children to research their habitats and create plants and animals from salt dough, art supplies, and collage materials. Suggestions for making items are as follows:

1. Use salt dough to create land features such as hills, rocks, and rivers. Air dry for a few days.

2. Form animals from salt dough, using broken toothpicks to attach animal parts. Add beads for eyes. Bake at 250°F until light brown and hard. Paint with tempera or acrylics.

3. Roll pieces of brown construction paper to make tree trunks, and cut leaves from green construction or tissue paper. Attach the trees by cutting slits in the bottom of the trunks and gluing the flaps to the base.

4. Add pebbles, stones, twigs, shells, moss, and other collage materials to complete the habitats.

LEARNING A SKILL

Drawing conclusions

A bright idea from Marlene Beierle and Anne Sylvan, multi-age teachers, Olde Orchard School, Columbus, Ohio

Children can make riddle books to share with friends by following these directions:

1. Fold a sheet of construction paper in thirds twice to make nine sections. Cut out each corner to make a cross shape.

Materials
✓ construction paper
✓ art supplies

2. Fold the top, bottom, and both sides into the middle.

3. Write a descriptive characteristic of a chosen animal on each folded section.

4. When all four folded sections are open, draw a picture of the animal in the center. Identify the animal with its name or a short sentence.

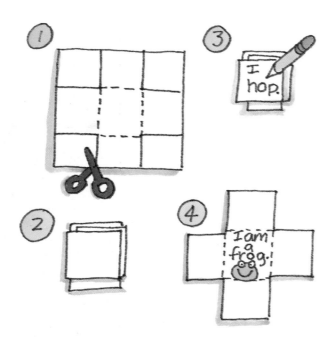

LINKING SCHOOL TO HOME

Research

Send home directions asking parents to help their child research an animal and its habitat. Suggest a family trip to the library. After learning about the animal, family members can help make a mobile. The mobile should prominently display the chosen animal with items hanging from it that represent what the animal needs to survive. Finished mobiles can be brought to school for display in the classroom.

Materials
✓ construction paper
✓ drawing paper
✓ yarn or string
✓ collage materials
✓ art supplies

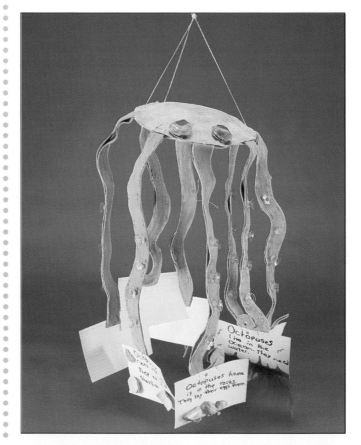

LITERATURE LINKS

And So They Build by Bert Kitchen

Animal Architecture by Jennifer Dewey

Animals and Where They Live by John Feltwell

Big Red Barn by Margaret Wise Brown

Cactus Hotel by Brenda Guiberson

Come Out Muskrats by Jim Arnosky

Hoot, Howl, Hiss by Michelle Koch

A House Is a House for Me by Mary Ann Hoberman

How Do Bears Sleep? by E.J. Bird

Large as Life by Julia Finzel

Otters Under Water by Jim Arnosky

Tree Trunk by Bianca Lavies

WE CAN EAT THE PLANTS

A child is followed by curious bunnies as he pushes a wheelbarrow through the garden, filling it with different fruits and vegetables.

FARMER'S MARKET
Three-dimensional mural

A bright idea and project from Majella Maas and her students, Lincoln School, Long Beach, California

Materials

✓ white butcher paper
✓ tempera paint
✓ paintbrushes
✓ newspaper
✓ art supplies

Refer to the big book version of *We Can Eat the Plants* and other related sources to make a list on the chalkboard of fruits and vegetables that come from different parts of a plant.

Pair children with partners. Have each pair choose an item from the list and make a "stuffed" fruit or vegetable. Start with two sheets of white butcher paper at least 12" square. Invite children to sketch the food item on the top sheet, making it as big as possible. Have them cut out their drawing with the other sheet underneath so they have two cutouts back-to-back. Staple around the edges, leaving a space open to fill the food item with crumpled newspaper, and finish stapling. Invite children to paint their food with tempera and attach it to the appropriate section of a painted backdrop.

Optional: Create an interactive display by attaching adhesive Velcro pieces to the food items and the background. Store the items in a bushel basket. Children will love attaching items to their correct place on the backdrop.

SCIENCE CONCEPTS

• Plant parts have names.
• People and animals eat plants.
• People eat different parts of different plants.
• People grow plants for food.

WRITING FRAMES

We can eat _____.

_____ comes from the _____ of a plant.

We can eat _____ but we can't eat _____.

People eat _____.
Animals eat _____.

RELATED SKILLS

• classifying
• picture details
• phonics: *ea (eat, leaves)*
• high-frequency words: *we, can, the*

STEMS AND LEAVES
Science experiments

The following experiments will lead to discussions on the function of a plant's stems and leaves:

Experiment 1: Fill both glasses with about one cup of water. Add twenty drops of red food coloring to one glass and twenty drops of blue food coloring to the other. Place a stalk of celery in each glass and observe for several days.

Note: The stalks and leaves of the celery will change color. Food and water travels to the leaves through the stems.

Materials
- ✓ celery stalks
- ✓ food coloring (blue, red)
- ✓ 2 clear glasses
- ✓ leafy houseplant
- ✓ petroleum jelly
- ✓ drawing paper
- ✓ paper clips

Experiment 2: Spread petroleum jelly on the bottom of a few leaves; spread petroleum jelly on the top of a few leaves; and cover a few leaves with paper. Observe the leaves for three or four days and record your findings.

Note: Plants need air. They get air through tiny holes on the underside of their leaves. The leaves with petroleum jelly on the bottom will die. Plants need sunlight, but the petroleum jelly does not block sunlight, so leaves with petroleum jelly on the top will not change. The leaves covered with paper will turn brown and die due to lack of sunshine.

LET'S EAT THE PLANTS
Tasting different plant parts

Help children clean, cut, and place food in a buffet arrangement. Place labels next to each item, identifying the plant and the part of the plant it comes from. Use ranch dressing and yogurt as dips and dressings. Children can sprinkle the seeds on the dressings.

While children are eating their snacks, discuss other plants they eat and the parts of the plants they come from. Have children draw and write about some of the plants they like to eat.

Materials
- ✓ carrots (roots), broccoli (flowers), celery (stem), lettuce (leaves), alfalfa sprouts (sprouts), bananas (fruit), sunflower seeds (seeds)
- ✓ ranch dressing
- ✓ yogurt
- ✓ paper plates
- ✓ plastic forks

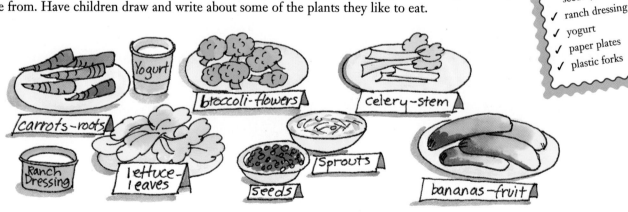

LEARNING A SKILL

Classifying

Materials

✓ seed catalogs
✓ index cards
✓ glue
✓ pocket chart

Invite children to glue pictures from seed catalogs to index cards to classify in a pocket chart.

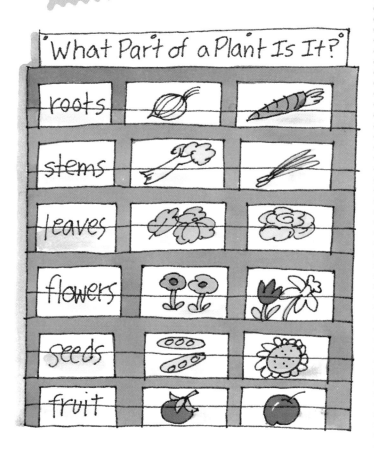

LINKING SCHOOL TO HOME

Food cards

Place several blank cards in a plastic bag along with the following directions:

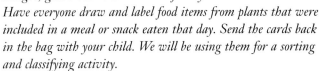

Materials

✓ index cards
✓ plastic zipper bags
✓ art supplies

Dear Parents,
We are currently learning about the parts of plants we eat. After dinner tonight, give a blank card to each family member. Have everyone draw and label food items from plants that were included in a meal or snack eaten that day. Send the cards back in the bag with your child. We will be using them for a sorting and classifying activity.

Add children's cards to the *Learning a Skill* activity described previously. Children will also enjoy sorting the cards according to different attributes.

LITERATURE LINKS

Apple Picking Time by Michele Benoit Slawson

The Biggest Pumpkin Ever by Steven Kroll

The Carrot Seed by Ruth Krauss

Cherries and Cherry Pits by Vera B. Williams

Eating the Alphabet: Fruits and Vegetables from A to Z by Lois Ehlert

Grandpa's Garden Lunch by Judith Casely

Growing Vegetable Soup by Lois Ehlert

Kimi and the Watermelon by Miriam Smith

Muskrat, Muskrat, Eat Your Peas! by Sarah Wilson

Peas by Nicholas Heller

Pumpkin, Pumpkin by Jeanne Titherington

Vegetable Garden by Douglas Florian

The Vegetable Show by Laurie Krasny Brown

THE SEED SONG

The story of what a pumpkin seed needs to grow is told through colorful pictures and a lively song.

SCIENCE CONCEPTS

• Plants are living things.

• A plant can grow from a seed.

• A plant grows from its own kind of seed.

• Plants need nutrients, air, light, and water to live and grow.

WRITING FRAMES

Seeds need _____ to grow.

A _____ grows from a _____ seed.

RELATED SKILLS

• sequencing

• phonics: *rhyming words (deep/sleep; bright/light; blow/grow)*

• punctuation: *commas*

*A*CTIVITY

PLANTS APLENTY
Classroom garden

Set up a gardening center near a sunny window in your classroom. Place the listed materials in the center along with easy-to-read direction cards. The following is a list of planting activities for young gardeners:

Materials
✓ potting soil
✓ clear plastic cups
✓ watering can
✓ cotton or paper towels
✓ small gardening tools
✓ seeds

1. Fill a clear plastic cup with moist cotton or paper towels and place bean seeds around the sides. (Fordhook lima beans work well because of their large size.) Place some of the beans upside down. Place the cup in a warm, dark place and keep the paper towels moist.

2. Seeds from oranges, grapefruits, and lemons grow into beautiful houseplants. Try growing seeds from apples, peppers, melon, or pears. Make sure to label the containers.

3. Try growing a "pixie" hybrid tomato as a houseplant. Place it in a sunny, south-facing window. Harvest your "pixie" tomatoes for a classroom snack.

4. Fill a tall, clear glass with water and place a sweet potato inside, supported by toothpicks. A green, leafy vine will grow.

5. Some plants will grow from cuttings. Ask children to bring in cuttings from houseplants at home. Place them in water and observe which ones grow roots.

6. Try growing a rainbow of flowers. Geraniums, impatiens, marigolds, and begonias all grow well in small spaces.

IT'S ALL IN THE BAG
Story bags

A bright idea and project from Marcia Fries and her son, Matthew, Los Alamitos, California

To make a story bag, have children decorate the front of their paper bags with coffee grounds, paint, lima beans, and construction paper scraps. On the back of the bags, glue the text from *The Seed Song*.

Give each child seven cards. Have children attach a line of text from the song to each card and illustrate it with art supplies and collage materials. Children can store the cards in the paper bag and take them out for a sequencing activity. They can refer to the song on the back of the bag if necessary.

Materials
✓ small brown paper bags
✓ large index cards
✓ coffee grounds
✓ dried lima beans
✓ art supplies
✓ collage materials
✓ copies of text from *The Seed Song*

SEE THE SEEDS
Hands-on activity

A bright idea from Judi Hechtman, first grade teacher, University School, Indiana, Pennsylvania

Materials
✓ variety of fresh fruits and vegetables with seeds
✓ paper plates
✓ potting soil
✓ small paper cups
✓ construction paper
✓ magnifying glasses

Have parents help supply fresh fruits and vegetables with seeds such as apples, citrus fruit, melons, pears, grapes, tomatoes, peppers, cucumbers, and green beans. You will need two of each fruit or vegetable.

Help children set up a display of the gathered foods. Make sure hands and work surfaces are clean before slicing one of each fruit and vegetable. Display each food on a paper plate so the seeds are visible. For identification purposes, place whole items next to those that are sliced and label accordingly. Make sure everyone gets a chance to examine the display. Have magnifying glasses handy. After viewing the display, wash the fruits and vegetables, remove the seeds, and let children snack.

Plant some of the seeds in paper cups, and label. Observe and chart the plant growth. Use the rest of the seeds to make little books, gluing a seed on a page and drawing the fruit or vegetable it came from.

ANOTHER SEED SONG
Writing lyric innovations

A bright idea and project from Debbie Martinez and her kindergartners, Helen Lehman School, Santa Rosa, California

Using the language pattern from *The Seed Song*, encourage children to write new lyrics for the song. Have children create a colorful flower mural, and attach sentence strips of the new verses.

Materials
- ✓ butcher paper
- ✓ construction paper
- ✓ art supplies
- ✓ collage materials

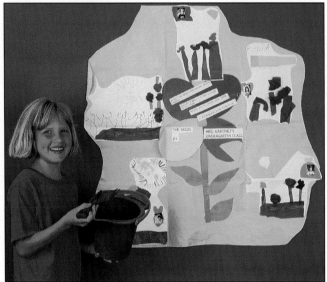

LEARNING A SKILL

Sequencing

A bright idea from Lori J. Avalos, first grade teacher, Los Alamitos Elementary, Los Alamitos, California

Discuss the importance of describing events sequentially in a story. After practicing the song, have children take turns placing the sentences in sequential order in the pocket chart.

Materials
- ✓ sentence strip text from *The Seed Song*
- ✓ construction paper
- ✓ art supplies
- ✓ pocket chart
- ✓ blue tagboard
- ✓ cloth tape
- ✓ clear plastic pockets

Children can make stand-up accordion books by illustrating the verses on sheets of blue tagboard. Attach the boards with cloth tape. Make the display interactive by adding clear plastic pockets for the sentence strips. Children can mix up the sentence strips and place them with the matching pictures.

LINKING SCHOOL TO HOME

Family sing-along

Have children take home their story bags. Children can share their work with family members and practice placing the cards in the correct sequence. Children will also enjoy teaching their family the song. Some might try accompanying the singing with an instrument. Suggest that families tape-record their renditions for the class to enjoy.

Materials
- ✓ paper bag stories from *It's All in the Bag* activity
- ✓ dried lima beans

Place a few lima beans in the take-home story bag. Encourage family members to plant the beans and chart their growth.

LITERATURE LINKS

Anna's Garden Songs by Mary Steele

A Flower Grows by Ken Robbins

From Seed to Plant by Gail Gibbons

Gardening for Kids by L. Patricia Kite

How a Seed Grows by Helene Jordan

Inch by Inch: The Garden Song by David Mallett

Jack's Garden by Henry Cole

More Than Just a Vegetable Garden by Dwight Kuhn

Planting a Rainbow by Lois Ehlert

The Pumpkin Patch by Elizabeth King

Ten Tall Oak Trees by Richard Edwards

The Tiny Seed by Eric Carle

SEE HOW IT GROWS

The stages of growth are shown for a tree, flower, butterfly, spider, bird, frog, and human.

See how the bird grows.   See how the frog grows.

See How It Grows

STAGES OF LIFE
The life cycle of a silkworm

Line the box or aquarium with mulberry leaves and place the silkworm eggs on top of the leaves. Tiny caterpillars will hatch. Have magnifying glasses near the setup for observing the tiny creatures. Place a science journal nearby where children can date and record observations. Add fresh mulberry leaves every two days or so.

Silkworms have voracious appetites and grow very quickly. They will not leave the container as other caterpillars might do and can be handled with care by children. After about three weeks, the round, fat caterpillars will spin cocoons. Place egg carton sections in the box so caterpillars will have something on which to attach their cocoons.

After about two weeks, moths will hatch. They do not fly and will not leave the box. They will mate, lay eggs, and die. The eggs can be refrigerated for next spring's hatching. The whole process is incredible to observe—children love the activity year after year.

Note: Silkworm eggs can be obtained from Insect Lore, P.O. Box 1535, Shafter, CA 93263 (1-800-548-3284).

Materials
- ✓ small aquarium or cardboard box
- ✓ silkworm eggs
- ✓ cardboard egg cartons
- ✓ mulberry leaves (must be available in your area)
- ✓ magnifying glasses
- ✓ science journal
- ✓ construction paper
- ✓ art supplies

SCIENCE CONCEPTS

- Living things grow and change.
- Living things can be classified as plants or animals.
- Living things have life cycles.
- Life cycles vary.

WRITING FRAMES

See how the _____ grows.

_____ grow from _____.

RELATED SKILLS

- sequencing
- classifying
- phonics: *ee (see)*
- high-frequency words: *see, how, the*
- parts of speech: *verbs (grow/grows)*

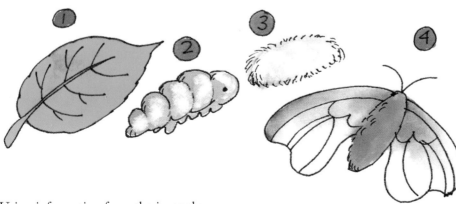

Using information from the journal, review the stages of growth with children. Have each child create a flip-book by folding a piece of 12" x 18" construction paper in half and cutting four flaps. On the top of each flap, children can number the stages one through four, or write a short sentence to describe each stage. Invite children to add illustrations underneath each flap.

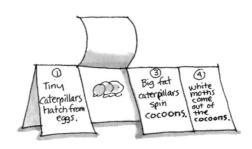

① Tiny Caterpillars hatch from eggs. ③ Big fat caterpillars spin cocoons. ④ White moths come out of the cocoons.

SEE HOW THE TEACHER GREW!
Sequencing stages of development

Place the photos in plastic bags and attach them in random order to a bulletin board. Add the number cards in order. Have children take turns arranging the pictures in the correct sequence.

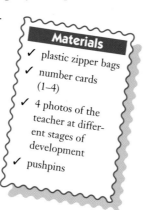

Materials
- ✓ plastic zipper bags
- ✓ number cards (1–4)
- ✓ 4 photos of the teacher at different stages of development
- ✓ pushpins

SEE WHAT GROWS
Matching game

Assign each group of three or four children one plant or animal that has been studied by the class. Invite each group member to draw and color the plant or animal at a different stage of development on an index card. All groups place their cards in one plastic bag.

Materials
- ✓ large index cards
- ✓ plastic zipper bags
- ✓ art supplies

Children can play the game in small groups. To play, have children divide the cards evenly and take turns creating the life cycles of plants and animals. The player that places the last card for any life cycle picks up those cards. The player with the most cards at the end wins.

YOU'LL SOON GROW!
Mobiles

Gather multiple copies of *You'll Soon Grow Into Them, Titch*. As you read the book together, have children carefully observe details in the illustrations and discover all the different living things that grow as the story unfolds (Titch, garden, houseplants, blossoming tree, and so on).

Materials
- ✓ *You'll Soon Grow Into Them, Titch* by Pat Hutchins
- ✓ index cards
- ✓ art supplies
- ✓ hole puncher
- ✓ string or yarn

Assign children to collaborative groups and ask each group to create a mobile that charts the growth of one living thing in the story or another living thing that the group has researched.

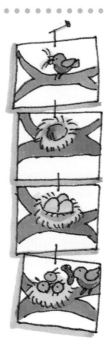

LEARNING A SKILL

Sequencing

A bright idea and project from Karen Bauer and her kindergartners, Vista Verde School, Irvine, California

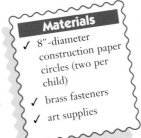

Materials
- ✓ 8"-diameter construction paper circles (two per child)
- ✓ brass fasteners
- ✓ art supplies

Have children choose animals from *See How It Grows* or research other living things and create a wheel book showing the living thing's growth in correct sequence. Connect the two circles with a brass fastener and cut a "window" from the top circle. Underneath, children can draw their living thing's life-cycle stages. Have children turn the top circle to see the stages in the correct sequence.

LINKING SCHOOL TO HOME

"See how we've grown" picture comparisons

Give each child two plastic bags with a piece of black construction paper inside. With help from parents, have children attach a baby picture to the black paper in one plastic bag and a recent picture in the other. Encourage parents and children to discuss the similarities and differences between the two pictures. Ask children to bring the pictures in the plastic bags to share with the class.

Materials
- ✓ plastic zipper bags
- ✓ black construction paper

Display all the baby pictures on one board and recent pictures on another. Children can try to match the pictures and play "Guess Who?"

LITERATURE LINKS

A Chick Hatches by Joanna Cole

The Caterpillar and the Polliwog by Jack Kent

Chick; Duck; Frog; Kitten; Puppy; Rabbit by Angela Royston

Chicken and Egg by Christine Back

Chickens Aren't the Only Ones by Ruth Heller

From Tadpole to Frog by Wendy Pfeffer

I'm Growing by Aliki

The Life Cycle of the Butterfly by Paula Hogan

The Life Cycle of the Frog by Paula Hogan

The Reason for a Flower by Ruth Heller

A Seed, a Flower, a Minute, an Hour by Joan Blos

A Seed Is a Promise by Claire Merrill

Silkworms by Sylvia A. Johnson

Too Many Chickens! by Paulette Bourgeois

You'll Soon Grow Into Them, Titch by Pat Hutchins

WHO'S HIDING?

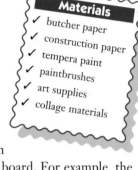

Who's hiding from the owl?   A mouse.

Close calls abound when animals in their natural habitats narrowly miss capture by predators.

SCIENCE CONCEPTS

• Animals have different structures that enhance the chance of survival in their environment.

• Living things are interrelated and dependent on each other.

• Nature is diverse.

• Animals live in different environments.

WRITING FRAMES

Who's hiding from the _____?

A _____ is hiding from the _____.

Come out, _____.

RELATED SKILLS

• question and answer format

• contraction: *who's*

• parts of speech: *verbs (hide/hiding)*

• punctuation: *question marks, commas*

• high-frequency words: *from, the*

LOOK WHO'S HIDING
Learning about animals and their habitats

A bright idea from Susan Gardner, teacher of multi-age class (K–3), Rolling Hills School, Fullerton, California

Arrange for an animal habitat tour at a local nature center, or take a walk around the neighborhood or school grounds. If this is not possible, "tour" animal habitats through the rich selection of literature available, including nature magazines such as *Big Backyard* or *Ranger Rick*. Decide which habitats you will focus on and list them on the board. For example, the class may decide on meadow, pond, ocean, and forest. Create three columns under each habitat labeled *Plants*, *Animals*, and *Other*. Have resource books available when creating the lists.

After gathering information, divide the class into groups and assign each group a habitat. Have each group create a painted mural of their habitat. They may add details using construction paper and flaps that can be lifted to reveal hidden animals.

Materials
✓ butcher paper
✓ construction paper
✓ tempera paint
✓ paintbrushes
✓ art supplies
✓ collage materials

Paintings by first graders, Lee School, Los Alamitos, California

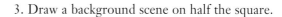

FOOD CHAINS
Quadraramas

A bright idea and project from Marcia Fries and her multi-age primary class, Lee School, Los Alamitos, California

Reread the big book version of *Who's Hiding?* focusing on the predator/prey relationship of animals. After learning about the food chain, children can make their own food chain quadraramas following these directions:

Materials
- ✓ 9" x 9" colored construction paper (4 sheets per child)
- ✓ art supplies
- ✓ craft materials

1. Fold down the top right corner of one construction paper square to the lower left corner. Repeat with the opposite corner.

2. Open and cut one fold line to the center of the square.

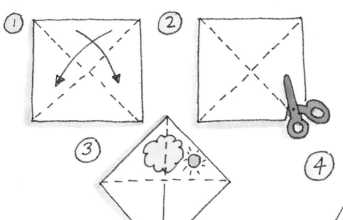

3. Draw a background scene on half the square.

4. Overlap the two bottom triangles and glue in place.

5. Make three more triaramas and glue them together to form a quadrarama.

6. Use art supplies and craft materials to depict four components of the food chain. Add text to each quadrant.

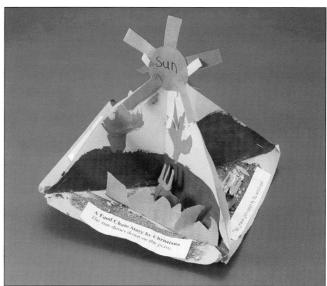

WHO LIVES HERE?
Matching

A bright idea from Susan Gardner, teacher of multi-age class (K–3), Rolling Hills School, Fullerton, California

Materials
- ✓ animal pictures
- ✓ construction paper
- ✓ science journals

Gather a variety of animal pictures. Nature magazines such as *Ranger Rick* and *Big Backyard* are good sources. Mount pictures on construction paper.

Distribute pictures to children working in small groups of three to five. Have children classify the pictures in a variety of ways such as: like environments or habitats; living habits (nocturnal vs. diurnal); and predators with prey. Children may exchange pictures with other groups. After each classification, have group members record the information in a science journal.

LEARNING A SKILL

Writing questions as dialogue

Children learn that writing dialogue is writing what characters actually say. Dialogue makes stories more interesting.

Invite children to use their knowledge of animals to create dialogue between the animals in *Who's Hiding?* After brainstorming with children, write dialogue on self-adhesive notes and stick them to the book pages for more reading fun.

LINKING SCHOOL TO HOME

Animal riddles

Have children take home a blank riddle sheet (see illustration). Parents can help children research one animal and write three clues about the animal on the left side of the sheet. Have children draw and label the animal on the right side and fold on the dotted line. Ask children to bring their finished sheets to school to be included in a class riddle book. All the pages should be folded so children can read the clues, guess the animal, and then unfold the paper to check their answers. Have children take turns bringing the book home to enjoy with family members.

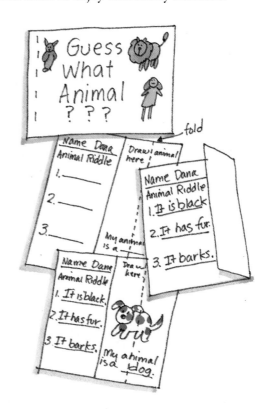

LITERATURE LINKS

All About Owls by Jim Arnosky

Animal Fact: Animal Fable by Seymour Simon

Animal Tracks by Arthur Dorros

The Architecture of Animals by Adrian Forsyth

Busy Beavers by Lydia Dabcovich

Can You Find Me? A Book About Animal Camouflage by Jennifer Dewey

The Icky Bug Alphabet Book by Jerry Pallotta

Ocean Animals (A Random House Tell-Me-About Book) by Michael Chirney

What's Hiding? by Mario Gomboli

Where's That Insect? by Barbara Brenner and Bernice Chardiet

Whoo-oo Is It? by Megan McDonald

THE FOUR SEASONS

Bold and colorful art depicts
things to see in each season.

It's summer.

I see flowers.

TIME TO TAKE A WALK
Seasonal changes mural

A bright idea and project from Marcia Fries
and her multi-age primary students, Lee
School, Los Alamitos, California

At least once each season, take a walk with your class around
the neighborhood or school grounds. Take along clipboards
to jot down things you see. After returning from the walk,
brainstorm a list of observations on the chalkboard.

Using their observations, have children paint a mural.
Sponge-paint the background first. Children can paint
details for the mural on drawing paper and cut them out
when the paint is dry. Attach sentence strips to the mural as a finishing touch.

Materials
✓ clipboards
✓ drawing paper
✓ butcher paper
✓ tempera paint
✓ paintbrushes
✓ small sponge squares
✓ sentence strips
✓ art supplies

SCIENCE CONCEPTS
• Seasons follow one another in a continuous pattern.

• Changes in nature can be observed during the four seasons.

WRITING FRAMES
It's _____. I see _____.

I see _____ in the _____.

I (<u>see, hear, etc.</u>) _____.

It must be (<u>season</u>).

RELATED SKILLS
• drawing conclusions

• contraction: *it's*

• phonics: *ee (see)*

PICTURE PERFECT
Observing seasonal changes

Take a picture of one designated area
in the neighborhood or on the school
grounds each season. Attach the labeled photo
to the class calendar. At the end of the school year, you'll
have all four seasons recorded with photography.

Materials
✓ camera

I SEE . . .
Seasonal characteristics display

A bright idea from Kathy Dahlin, first grade teacher, Cooper School, Superior, Wisconsin

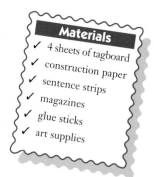
Divide the class into four cooperative groups. Assign each group one of the four seasons. Give each group one sheet of tagboard. Ask groups to fill their tagboard with pictures cut from magazines that tell about their season. Children can use construction paper to make things they do not find in magazines.

When each group has finished, have them trade tagboards. Give each group member a sentence strip. Have children look at the tagboard and write a sentence describing one thing they see, such as "I see mittens." Encourage them to find different things. Keep tagboards displayed so each group has a chance to work on every season.

Place the finished tagboards and sentence strips at a center. Children can take turns reading the sentence strips and matching them to pictures on the tagboards.

WHAT DO YOU SEE?
Four seasons mural

A bright idea and project from Gerianne Smith and her kindergartners, Minnie Gant School, Long Beach, California

Discuss the four seasons with children. Divide the chalkboard into four sections and label them *spring*, *summer*, *fall*, and *winter*. Ask children to brainstorm typical things they observe during each season. List these in the appropriate sections. Have each child choose one thing from the board to draw, color, and outline with black marker. Children can use sentence strips with the writing frame "I see _____" to describe their pictures. Cut out and attach the pictures and sentence strips to a blue butcher paper background.

SEASONAL WRITES

Seasonal postcards

Give each child four tagboard "postcards." On one side, have children draw and color seasonal scenes—one for each season or four scenes of the current season. On the back of each card, instruct children to draw a vertical line a little to the right of the center. Explain that the address goes to the right of the line and the message goes on the left. Children can write or dictate their own messages or use the frame: "It's _____. I see _____." Help children address their postcards.

Materials
- ✓ art supplies
- ✓ tagboard (postcard size)
- ✓ pen pals (optional)

Encourage children to mail postcards to friends, relatives, or pen pals, or use them to write positive messages to children's families throughout the year.

LEARNING A SKILL

Drawing conclusions

Read passages from different books that clearly describe seasonal characteristics. A good choice is *Frog and Toad All Year.* After reading a paragraph, have children conclude which season was described and identify phrases that support their answers.

Materials
- ✓ books with seasonal themes

LINKING SCHOOL TO HOME

Seasonal boxes

Encourage parents to help children gather items relating to the current season. Items can be placed in a box or bag and brought to school for sharing. Invite children to give reasons why they chose the items. When everyone has shared, discuss the similarities and differences of everyone's choices.

Materials
- ✓ small boxes or bags
- ✓ various seasonal items

LITERATURE LINKS

A Busy Year by Leo Lionni

All Year Long by Nancy Tafuri

Caps, Hats, Socks, and Mittens by Louise Borden

The Chipmunk's Song by Joanne Ryder

Fall; Winter; Spring; Summer by Ron Hirschi

Four Stories of the Four Seasons by Tomie dePaola

Red Leaf, Yellow Leaf by Lois Ehlert

The Snowy Day by Ezra Jack Keats

Songs for the Season by Jamake Highwater

Why Do Leaves Change Color? by Betsy Maestro

ROUND AND ROUND THE SEASONS GO

An easy-to-learn song highlighting the cyclical nature of the seasons.

SCIENCE CONCEPTS

• Seasons follow one another in a continuous pattern.

• Weather changes with the seasons.

• Nature changes with the seasons.

WRITING FRAME

Round and round the seasons go.

_____ comes, _____ _____.

RELATED SKILLS

• picture details

• phonics: *rhyming words (go, snow, grow, slow, blow)*

• parts of speech: *descriptive words (adjectives)*

• classifying

• punctuation: *commas*

ACTIVITY

THE SEASONS OF OUR TREE
Interactive mural

Use the brown butcher paper to "paper sculpt" a 3-D tree trunk and branches on a blue background of a large bulletin board space. Divide the tree into four quadrants and have children decorate each of the sections to represent the tree in each of the four seasons.

For winter, use white paper to make snow and snowflakes; for spring, use light green paper for leaves and pink tissue for blossoms; for summer, use dark green paper to make leaves; and for autumn, use yellow, red, and orange construction paper to make fall leaves. Make other details, such as flowers, birds, or butterflies, from construction paper scraps and attach to the display.

Add text from *Round and Round the Seasons Go* on sentence strips to make the display a great stopping point for children when "reading the room."

Materials
✓ butcher paper (brown, blue)
✓ construction paper
✓ pink tissue paper
✓ sentence strips
✓ art supplies

I KNOW WHAT SEASON IT IS
Class book

A bright idea from Renee Keeler, first grade teacher, Lee School, Los Alamitos, California

Give each pair of children a page for a class book cut from butcher paper with the following frame written on it:

Materials
✓ blue butcher paper
✓ art supplies
✓ collage materials

_____ *is here,*
_____ *is here.*
How do you think I know?
I (saw, heard, felt, etc.) _____.
I know it must be so!

Have partners complete the writing frame together and use art supplies and collage materials to illustrate their page according to the current season.

Combine the pages for a class book. Have children create a tune to go along with the text.

CIRCLE OF SEASONS
Wheel books

A bright idea and project from Marcia Fries and her multi-age primary students, Lee School, Los Alamitos, California

Materials
✓ 12"-diameter tagboard circles (two per child)
✓ brass fasteners
✓ art supplies
✓ text from *Round and Round the Seasons Go* reproduced in 18-pt. font size

Have children make their own wheel books using reproduced text from *Round and Round the Seasons Go*. Begin by attaching the tagboard circles with a brass fastener. Help children cut a "window" from the top circle. After decorating the top circle and adding the title, have children add text and seasonal art to each of the four quadrants of the bottom circle. When completed, turn the top circle to reveal lyrics and art in the "window."

WHAT DO WE WEAR?
Class chart

Create a chart with columns labeled *Winter, Spring, Summer,* and *Fall.* Label the rows *jacket, sweater, mittens, hat, long pants, shorts, long sleeves, short sleeves, bathing suit,* and so on.

Materials
✓ chart paper or butcher paper
✓ clothing catalogs

Have children look through copies of *Round and Round the Seasons Go*. Invite volunteers to check the box showing the season when each article of clothing is usually worn. As an extension, have children look through catalogs to find pictures of clothing to cut out and add to the appropriate sections of the chart.

Clothing	Winter	Spring	Summer	Fall
jacket	✓			✓
sweater	✓			✓
mittens	✓			
hat	✓	✓	✓	✓
long pants	✓			✓
short pants		✓	✓	
long sleeves	✓			✓
short sleeves		✓	✓	
bathing suit			✓	
boots	✓			✓

LEARNING A SKILL

Picture details

Have children look for details in pictures from *Round and Round the Seasons Go* that unfold as little stories of their own. For example, look for the cat in all the pictures or the bird building its nest. Encourage children to describe or write "subtexts" gleaned from picture details.

Materials

✓ big book version of *Round and Round the Seasons Go*

LINKING SCHOOL TO HOME

Take-home book

Have children take their *Circle of Seasons* books home to share with family members. What a great opportunity for rereading and sharing their science knowledge!

Materials

✓ wheel book from *Circle of Seasons* activity

LITERATURE LINKS

A Year in the Country by Douglas Florian

Chicken Soup with Rice by Maurice Sendak

First Comes Spring by Anne Rockwell

Frog and Toad All Year by Arnold Lobel

The Goodnight Circle by Carolyn Lesser

January Brings the Snow by Sara Coleridge

My House by Lisa Desmini

The Seasons of Arnold's Apple Tree by Gail Gibbons

Sky Tree: Seeing Science Through Art by Thomas Locker

Spring Is Here by Taro Gomi

Sunshine Makes the Seasons by Franklyn Branley

Voices on the Wind: Poems for All Seasons edited by David Booth

HOW'S THE WEATHER?

A fun-to-read book describing
different kinds of weather.

It's a cloudy day.   It's a windy day.

A*ctivity* WHAT'S BLOWIN' IN THE WIND?
Finger painting

Read the book *The Wind Blew* and/or learn the song "The Wind" (see Literature Links). The repetitive language in the verses of the song makes a great frame for children's own writing:

Who's that blowin' blossom seeds in the air?
Who's that blowin' willow fluff in my hair?
The wind, the wind, the wind, the wind!

Give each child a piece of finger-painting paper and pour a few tablespoons of liquid starch on it. Have children follow the rules of finger painting by using only one hand to paint, keeping the other hand free to get supplies or scratch a nose. After children spread the liquid starch over the paper, squeeze on some blue paint. Children can create a design that looks like the wind. When finished, have children wash their hands and use torn paper to create objects blowing in the wind. Have them place these on their finger paintings before the paint dries.

Children can write new lyrics for "The Wind" and attach sentence strips to their pictures. Compile all pages into a class book or display as a wall story.

Materials
✓ finger-paint paper or glossy paper
✓ liquid starch
✓ liquid tempera paint (shades of blue)
✓ construction paper scraps
✓ *The Wind Blew* by Pat Hutchins
✓ "The Wind" by Jack Grunsky

SCIENCE CONCEPTS

• Weather can be observed.

• Weather changes from day to day, place to place, and season to season.

• People make decisions about clothing based on the weather.

WRITING FRAMES

It's a _____ day.

It's a _____ day. I will wear _____.

Let's go out and _____.

The weather is _____.

RELATED SKILLS

• picture details

• weather words

• parts of speech: *descriptive words (adjectives)*

• contractions: *it's, let's*

• phonics: *rhyming words (day/play)*

CLOTHES THERMOMETER
Manipulative temperature charts

A bright idea from Harri Wasch, kindergarten teacher, St. Anne's-Belfield School, Earlysville, Virginia

Reproduce a copy of the thermometer (as shown) for each child on 8 1/2" x 11" tagboard. Have children label the thermometer from 10°F to 100°F, and cut slits at the top and bottom.

Tape the ends of the red and white ribbon together and run it through the slits so it can be pulled up or down to indicate temperature. Once the ribbon is threaded, connect the ends in back so it won't pull out.

Invite children to look through magazines or clothing catalogs and cut out of different types of clothing. Discuss the meaning of the numbers on the thermometer and connect the differences in temperature to the types of clothing worn. Children can then glue their pictures to the appropriate section of the thermometer.

Materials
- ✓ art supplies
- ✓ clothing catalogs and magazines
- ✓ glue sticks
- ✓ tagboard
- ✓ 1/4"-wide, red and white ribbon

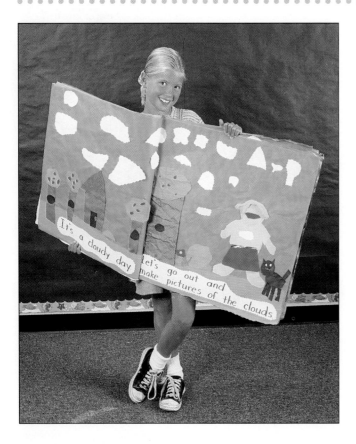

LET'S CHECK OUT THE WEATHER
Class big book

A bright idea and project from Marcia Fries and her multi-age primary students, Lee School, Los Alamitos, California

Supply children with the writing frame:

It's a _____ day. Let's go out and _____.

Invite children to brainstorm a list of new sentence possibilities and decide as a group which sentences to use in a class book. Attach the new sentence strips to big book pages. Have small groups of children illustrate them using cut paper and collage materials.

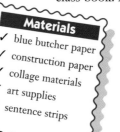

Materials
- ✓ blue butcher paper
- ✓ construction paper
- ✓ collage materials
- ✓ art supplies
- ✓ sentence strips

LEARNING A SKILL

Picture details
Action words (verbs)

Read the big book version of *How's the Weather?* with children, and discuss each picture's details. Have children create sentences describing what's happening to Red Bird and Cat on each page. Transcribe their sentences onto sentence strips, writing the verbs with a different color marker. Place the sentence strips in a pocket chart and use them for a lesson on action words. Children can also create illustrations for the pocket chart.

Materials
✓ big book version of *How's the Weather?*
✓ sentence strips
✓ art supplies
✓ pocket chart

LINKING SCHOOL TO HOME

Take-home books

A bright idea and project from Jennifer Botenhagen, Marcia Smith, and their first graders, Lu Sutton School, Novato, California

Children will love to make and take home their own illustrated versions of the book to share with family members. Include a "comments" page that can be filled out and returned to school.

Materials
✓ drawing paper
✓ art supplies

LITERATURE LINKS

Feel the Wind by Arthur Dorros

Froggy Gets Dressed by Jonathan London

Geraldine's Snow by Holly Keller

The Jacket I Wear in the Snow by Shirley Neitzel

Rain by Robert Kalan

Sun Up, Sun Down by Gail Gibbons

Weather by Pascale de Bourgoing

Weather Experiments (A New True Book) by Vera Webster

Weather Forecasting by Gail Gibbons

Weather Words and What They Mean by Gail Gibbons

The Wind Blew by Pat Hutchins

MUSIC

"The Wind" by Jack Grunsky from *Dream Catcher* (Youngheart Music)

WHAT'S THE WEATHER LIKE TODAY?

A fun-to-sing song that explores different weather conditions.

A*CTIVITY* OUR OWN BOOK
Cut paper art project

A bright idea and project from Rosa Drew and her first graders, Vista Verde School, Irvine, California

Invite children to experiment with the cut paper art technique used by the book's illustrator. Using the listed materials, have children work in small groups to illustrate their own version of *What's the Weather Like Today?*

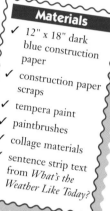

Materials
✓ 12" x 18" dark blue construction paper
✓ construction paper scraps
✓ tempera paint
✓ paintbrushes
✓ collage materials
✓ sentence strip text from *What's the Weather Like Today?*

A*CTIVITY* HOW MUCH RAIN?
Making a rain gauge

Rinse out the plastic bottle and remove label. With sharp scissors, cut off the top of the bottle about 8" from the base. Invert the top of the bottle, insert it into the base, and secure with plastic tape. Invite children to decorate the outside of the bottle and mark inches and half inches. Set the rain gauge in a box of sand outdoors. After each rainfall, measure the water and add the measurement to a cumulative class graph.

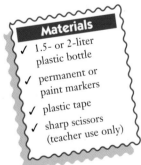

Materials
✓ 1.5- or 2-liter plastic bottle
✓ permanent or paint markers
✓ plastic tape
✓ sharp scissors (teacher use only)

WEATHER WIZARD
Class weather report and graph

A bright idea from Kim Jordano, kindergarten teacher, Lee School, Los Alamitos, California

A few mornings a week, have children take turns being the Weather Wizard and report on the condition of the day's weather to the class. The Weather Wizard can dress up with the cape, hat, and magic wand. Invite everybody to sing the chorus:

What's the weather like today, like today, like today? What's the weather like today?

The Weather Wizard can sing the last line:

Today is windy!

The Weather Wizard can turn the hand on a bulletin board weather wheel, look at an outdoor thermometer to report the temperature, and add a weather symbol to the calendar or monthly weather graph.

Materials
✓ magician's cape
✓ magic wand
✓ wizard's hat

WEATHER SONG
Group activity

A bright idea from Mary Kurth, kindergarten teacher, Black Earth School, Black Earth, Wisconsin

Have children draw or paint pictures on tagboard to make six weather condition cards. Add weather words to each card.

Each day, invite six children to sit in front of the class and hold the cards. As the class sings the lyrics to *What's the Weather Like Today?*, one child at a time stands and shows his or her weather card.

After the song, recite the following lyrics according to the weather conditions:

Is today sunny? No! No! No!
Is today windy? No! No! No!
Is today foggy? No! No! No!
Is today cloudy? Yes! Yes! Yes!
Is today rainy? Yes! Yes! Yes!

Materials
✓ six 8" x 12" pieces of tagboard
✓ art supplies

LEARNING A SKILL

Descriptive words (adjectives)

List weather conditions on the chalkboard under headings such as *A Rainy Day, A Windy Day,* or *A Sunny Day.* Discuss describing words with children and brainstorm appropriate adjectives describing each weather condition. Children can refer to the lists when writing their own weather poems using the frame in the illustration. Have them exchange poems with partners and underline describing words in each other's work.

Materials
- ✓ drawing paper
- ✓ art supplies
- ✓ chalkboard

LINKING SCHOOL TO HOME

Weather watch

Send a note home asking parents to watch a weather report on TV with their children or find one in the newspaper. After discussing the report, suggest that family members choose clothes for the next day based on the weather predictions. Encourage newspaper readers to bring in weather maps to share.

Materials
- ✓ newspaper or TV weather report

LITERATURE LINKS

Caps, Hats, Socks, and Mittens by Louise Borden

The Cloud Book by Tomie dePaola

Cloudy with a Chance of Meatballs by Judi Barrett

Dry or Wet? by Bruce McMillan

Gilberto and the Wind by Marie Hall Ets

Is It Dark? Is It Light? by Mary Lankford

It Looked Like Spilt Milk by Charles G. Shaw

The Sun's Day by Mordicai Gerstein

The Wind Blew by Pat Hutchins

WHOSE FOREST IS IT?

Forest inhabitants realize that the forest belongs to them all.

"This is *my* forest," said the fox.

"This is *my* forest," said the rabbit.

ACTIVITY

WHOSE OCEAN IS IT?

Three-dimensional interactive display

A bright idea and project from Trisha Callella and her kindergartners, Rossmoor School, Los Alamitos, California

Refer to the language pattern in *Whose Forest Is It?* to create a wall story about the ocean habitat. Brainstorm with children a list of creatures who live in the sea. Have partners choose an animal from the list to paint on white construction paper and cut out.

Give partners a tagboard sentence strip about their animal which they can attach to the "sand" (tan construction paper cut to resemble the ocean floor).

Materials
- ✓ blue butcher paper
- ✓ tagboard strips
- ✓ white construction paper
- ✓ tempera paint
- ✓ paintbrushes
- ✓ 12" x 18" tan construction paper (cut in half)
- ✓ self-adhesive Velcro
- ✓ beach buckets

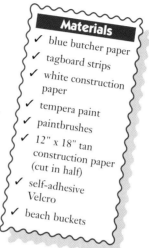

To assemble the wall story, follow these directions:

1. Cut a long strip of blue butcher paper about 2 1/2' wide.

2. Cut the top and bottom edges to resemble ocean waves.

3. Accordion-fold the paper into pages about 18" wide and open up again.

4. Fold the bottom edge up twice to make two folds about 3" apart. Make the bottom fold about 10" from the bottom edge.

5. Glue tagboard strips on each page between the folds.

6. When hanging the wall story, fold the bottom section up for a 3-D look.

7. Attach Velcro pieces to the sea animals and butcher paper display. Keep the animals in beach buckets. Have children read the text and find the right animal for each section.

SCIENCE CONCEPTS

- The forest is one of the earth's natural resources.
- The earth's resources need to be protected and preserved.
- People and animals depend on their distinctive habitats for life.
- Habitats must be preserved.
- A variety of animals share a common habitat.

WRITING FRAMES

"This is my _____," said the _____.

The forest belongs to _____.

A _____ lives in the _____.

RELATED SKILLS

- punctuation: *quotation marks, question marks*
- italics
- possessives

ACTIVITY

LET'S PUT ON A PLAY!
Creative drama

A bright idea and project from Trisha Callella and her summer school class, Rossmoor School, Los Alamitos, California

After reading and reviewing *Whose Forest Is It?* and other related books, help children brainstorm a list of forest animals and record them on a chart. Each child can choose an animal from the list and create a mask.

Materials
- ✓ tagboard pieces
- ✓ tongue depressors
- ✓ construction paper scraps
- ✓ blue butcher paper
- ✓ tempera paint
- ✓ paintbrushes
- ✓ art supplies

Each child can be a spokesperson for his or her animal. Have children start with the line: *This is **my** forest.* They then add one or two more sentences about how or why the forest habitat should be preserved.

After everyone has practiced their lines, paint a forest background mural on butcher paper and present the play to another class. After each "animal" has spoken, have an owl character say, *Who-o-o-o-se forest is it?* Everyone can answer, *It's **everyone's** forest!*

ACTIVITY

DAY AND NIGHT
Nocturnal and diurnal animal mobiles

Read the big book version of *Whose Forest Is It?* and discuss with children the nocturnal and diurnal animals appearing in the book. List these on a chart and add more forest animals to each column.

Materials
- ✓ paper plates
- ✓ crayons
- ✓ watercolors or watered-down tempera paint (blue, black)
- ✓ paintbrushes
- ✓ big book version of *Whose Forest Is It?*

Give each child a paper plate. On one side of their paper plates, have children draw and color a daytime scene in the forest, including diurnal animals. When the picture is completely colored in, have children paint the scene with a wash of blue paint. When this is dry, repeat the process on the back of the paper plate with a nighttime scene, including nocturnal animals. Cover this side with a wash of black paint. Label the scenes *Daytime* and *Nighttime*, and hang the plates as mobiles.

LEARNING A SKILL

Possessives

Reread the book with children, highlighting the use of the possessive pronoun *my* and how it's used in the text. Discuss how words can be changed to show ownership by adding *'s*. With

Materials
✓ self-adhesive notes

children, rewrite the text of the book using this frame: "It's the bear's forest," "It's the fox's forest," and so on. Write sentences on self-adhesive notes and attach them to the book pages.

LINKING SCHOOL TO HOME

This is my neighborhood!

Send blank booklets home with children that contain the following writing frame:

Materials
✓ blank booklets
✓ art supplies

"This is my _____," said the _____.

Ask parents and children to take a stroll in the backyard or neighborhood and complete the book with elements of the mini-environment they chose. Children can bring completed books to school to share with classmates.

LITERATURE LINKS

A Tree Is a Forest by Jan Thornhill

Baby Night Owl by Leslie McQuier

How the Forest Grew by William Jaspersohn

The Midnight Farm by Reeve Lindberg

Mighty Tree by Dick Gackenbach

Mockingbird Morning by Joanne Ryder

My River by Shari Halpern

Nature Walk by Douglas Florian

Over in the Meadow by Ezra Jack Keats

The Sun's Day by Mordicai Gerstein

Walk a Green Path by Betsy Lewin

We're Going on a Bear Hunt by Michael Rosen

What the Moon Saw by Brian Wildsmith

LET'S TAKE CARE OF THE EARTH

Six of Earth's habitats and animals who live there are showcased.

A͟CTIVITY THINK OF THE EARTH'S FUTURE
Bulletin board display

A bright idea from Trisha Callella, first grade teacher, Rossmoor School, Los Alamitos, California

Materials
✓ butcher paper (blue, white, brown)
✓ heart-shaped writing paper
✓ tempera paint (blue, green)
✓ paintbrushes
✓ newspaper
✓ art supplies
✓ collage materials

Cover a bulletin board with blue butcher paper. Add the title *The Future of the Earth Is in Our Hands*. Cut out a large circle from white butcher paper and draw the continents. Outline the continents with black marker and have volunteers paint the continents green and the oceans blue. Staple the earth shape to the bulletin board and add stuffing (crumpled newspaper) for a 3-D effect. Have each child make a handprint using green or blue paint on white paper. When dry, cut out the handprints and attach them around the earth shape.

Have each child make a self-portrait paper doll using brown butcher paper, art supplies, and collage materials. Children can attach button "eyes," yarn "hair," felt clothes, and so on. Invite children to write about ways to protect the earth on heart-shaped writing paper. Attach the paper dolls and hearts to finish the display.

EARTHKEEPER BACKPACKS
Paper bag backpacks

Help children make backpacks by following these directions:

1. Turn the paper bag so the flap that forms the bottom is on top and closest to you.

2. Round off the top corners of the flap and the top of the bag with a pair of scissors.

3. Staple lengths of wide ribbon to form straps.

Materials
✓ large brown grocery bags (one per child)
✓ magazines
✓ art supplies
✓ wide ribbon

Children can choose an animal and habitat from the book or class study. Have them decorate the outside of the backpack to represent the habitat. Invite them to fill their backpack with pictures of their chosen animal and what it needs to survive. They may draw pictures or find them in magazines.

Have children lift up the bottom flap of the paper bag and turn it under to form a clear writing surface. Using the frame, children can fill in the blanks with words relating to their chosen habitat.

The <u>desert</u>, the <u>desert</u> is home to a <u>coyote</u>. Let's take care of the <u>desert</u>.

As an oral presentation, each child can share his or her backpack and its contents.

TAKE CARE!
Class big book

A bright idea and project from Shawn Larson and her second graders, Cooper School, Superior, Wisconsin

Use the pictures and writing frame from *Let's Take Care of the Earth* as a starting point to extend children's knowledge of habitats. Choose one habitat from the book to study in depth. As part of your study, create a class big book that highlights the plants, animals, and other things that belong in the habitat.

Materials
✓ tagboard
✓ art supplies

LET'S DO IT!

Class environmental project

With children, discuss adopting "a small piece of the earth" as a service project.

This may be a small area on school grounds or in the community. For example, you may choose an area near the school office or a part of the playground. The class could plant and take care of flowers and keep the area free of litter. At the end of the school year, the class may want to "will" their area to the next year's class.

LEARNING A SKILL

Contractions

After discussing the contraction *let's*, have children write "let us" on self-adhesive notes. Attach the notes to the book pages and reread the book together, discussing which version everyone prefers. Review other contractions and make contraction fold cards. Add to the set as new contractions are introduced.

Materials
✓ self-adhesive notes

LINKING SCHOOL TO HOME

Backpack sharing

Have children wear their backpacks home and share the contents with family members. Encourage parents and children to look through current newspapers or magazines to find articles concerning local environmental issues. Post them on a bulletin board entitled *Earthkeepers Aware*. Share and discuss articles with the class.

Materials
✓ backpacks from *Earthkeeper Backpacks* activity
✓ current newspapers or magazines

LITERATURE LINKS

A River Wild by Lynne Cherry

Come Back, Salmon by Molly Cone

The Great Kapok Tree: A Tale of the Amazon Rain Forest by Lynne Cherry

Just a Dream by Chris Van Allsburg

Oil Spill! by Melvin Berger

Prince William by Gloria Rand

Tiger by Judy Allen

V for Vanishing: An Alphabet of Endangered Animals by Patricia Mullins

The Wartville Wizard by Don Madden

Welcome to the Green House by Jane Yolen

REDUCE, REUSE, RECYCLE

A short, repetitive song with illustrations of children reducing, reusing, and recycling common materials.

 THINK EARTH!
Writing a song

A bright idea and project from Marcia Fries and her first graders, Hopkinson School, Los Alamitos, California

Materials

✓ blue butcher paper
✓ sponge-painting supplies
✓ art supplies
✓ collage materials

Sing the following refrain to the tune of "Twinkle, Twinkle Little Star":

Think Earth, think Earth,
Is our song.
Everyone should sing along!*

After brainstorming a list of "Earth-saving" ideas, help children write lyrics for the song. For example:

Don't waste paper. Save a tree.
Turn off lights and the TV.

When the song is complete, illustrate the lyrics in a class big book.

SCIENCE CONCEPTS

• All resources used by humans—including fuels, metals, and building materials—ultimately come from the earth.

• Resources must be used with care, conserved, and recycled.

WRITING FRAMES

Your _____ is my _____, and my _____ is your _____.

Recycle by _____.

Reuse by _____.

Reduce by _____.

RELATED SKILLS

• picture details

• prefix: *re-*

• punctuation: *commas*

• contraction: *it's*

---

*From the *Think Earth* Environmental Education Program, © 1991 by Educational Development Specialists, Lakewood, California.

THE THREE Rs
Flip-book

A bright idea from Rosa Drew, first grade teacher, Vista Verde School, Irvine, California

Fold paper in half vertically and then in thirds horizontally. Open the paper and place it in front of you. Cut along the two parallel, vertical folds until you reach the fold at the center of the paper. Fold the flaps down.

Have children label the three flaps with the words *Reduce*, *Reuse*, and *Recycle*. Ask children to open the flaps and draw pictures showing things they could do to reduce, reuse, and recycle.

SMART ART
Recycled art show

For about a week, have a "junk drive" by asking children to bring in reusable items such as tissue rolls, egg cartons, paper scraps, cans, bottles, and plastics. When enough "junk" has been collected, have children create their own works of art from the materials. Display the creations "museum-style" adding titles and artists' names. Invite other classes to view the exhibit.

NO-TRASH LUNCH!
Classroom ecology

One day before lunch, ask children to save all of their non-organic trash from lunch in a trash bag. Two children can volunteer to be trash collectors. All paper, cardboard, metal (aluminum foil) and paper items should be collected. After lunch, weigh the bag. Discuss how most or all of the trash could be eliminated. Plan a "No-Trash Lunch." List ways children can bring lunches using reusable items. Suggest that everyone use canvas lunch bags, plastic containers for food and beverages, and cloth instead of paper napkins. After the "No-Trash Lunch," collect any waste and compare it to the previous day's waste. Encourage children to adopt some "Earth-saving" practices from the special lunch.

LEARNING A SKILL

Picture details

Use the big book version of *Reduce, Reuse, Recycle* and "read" the pictures. Look for information to list on a chart under the headings *Ways to Reduce*, *Ways to Reuse*, and *Ways to Recycle*.

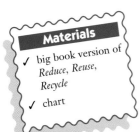

Materials
✓ big book version of *Reduce, Reuse, Recycle*
✓ chart

LINKING SCHOOL TO HOME

Sharing ideas

Send home copies of *Reduce, Reuse, Recycle* and/or children's flip-books. Ask family members to brainstorm ways they can reduce, reuse, or recycle at home. Have them create a recycling plan. Children can bring this information to school to add to the class chart described in the *Learning a Skill* activity.

Materials
✓ copies of *Reduce, Reuse, Recycle*
✓ flip-books from *The Three Rs* activity

LITERATURE LINKS

City Green by Dyanne DiSalvo-Ryan

Come Back, Salmon by Molly Cone

The Great Trash Bash by Loreen Leedy

Just a Dream by Chris Van Allsburg

Recycle! A Handbook for Kids by Gail Gibbons

Roxaboxen by Barbara Cooney

Trash! by Charlotte Wilcox

The Wartville Wizard by Don Madden

Where the Forest Meets the Sea by Jeannie Baker

Window by Jeannie Baker

IF A TREE COULD TALK

The message to protect the environment is related through rhyming verse and colorful cut paper art.

SCIENCE CONCEPTS

• People need clean air, clean water, and many kinds of plants and animals.

• People must protect the earth's resources.

• Human exploitation has damaged the earth's resources.

WRITING FRAMES

If a _____ could talk, what would it say?

"Don't _____, just walk away."

RELATED SKILLS

• punctuation: *quotation marks, question marks*

• contraction: *don't*

• phonics: *rhyming words (say, away, stay, day)*

• making inferences

• question and answer format

A**CTIVITY** THE PLAY'S THE THING
Creative drama

List "parts" of the environment that need protection, starting with those appearing in the book. Divide children into collaborative groups of three or four. Each group can choose something from the list to represent in a class play. They can decide as a group how they will depict their part of the environment using the materials available. For example, if the group is portraying animals, they may make masks or headbands. Another group could paint a river on a length of butcher paper.

Invite groups to use the frame from the book to write their own lines. Choose a narrator to read the introductory question, while group members respond with their line. After some practice, perform the play for parents or another class.

Materials
✓ art supplies
✓ collage materials

WHAT'S AN OIL SPILL?

Hands-on experiment

Fill the soda bottle half full with water (colored blue) and half full with mineral oil. Cap the bottle tightly and turn it on its side. Gently rock the bottle back and forth and note the results. Place a feather in the bottle and let children shake and rock it. After a few days, have children retrieve the feather and examine it. Discuss the effects that polluted water may have on birds.

Materials
- ✓ large clear plastic soda bottle with cap
- ✓ water
- ✓ blue food coloring
- ✓ mineral oil
- ✓ feathers

WHAT ELSE WOULD A TREE SAY?

A bright idea and project from Majella Maas and her students, Lincoln School, Long Beach, California.

Read *A Tree Is Nice* to the class. Make a list of all the good things about trees. Have children do easel paintings of a favorite tree, maybe one near the school or in their yards. When dry, have children cut out their paintings.

Materials
- ✓ *A Tree Is Nice* by Janice May Udry
- ✓ butcher paper
- ✓ tempera paint
- ✓ paintbrushes
- ✓ easels
- ✓ art supplies
- ✓ sentence strips

Using the list of good things about trees, have each child write or dictate on sentence strips something else a tree might say. Attach trees and sentence strips to the mural.

LEARNING A SKILL

Quotation marks

Invite children to help write the text from the book on sentence strips. They can glue elbow macaroni to represent quotation marks around the dialogue. Place the completed sentence strips in a pocket chart for a learning center or shared reading. Children can draw and color simple pictures to match the sentences in the pocket chart.

Materials
- ✓ sentence strips
- ✓ dry elbow macaroni
- ✓ pocket chart
- ✓ art supplies

LINKING SCHOOL TO HOME

Protecting our neighborhoods

To extend practice using quotation marks, send home two sentence strips and macaroni with each child. Ask parents and children to investigate their home or neighborhood and write about something that needs protection. Children can show parents what they know about quotation marks by gluing macaroni in the correct places. Have children bring sentence strips to school for the pocket chart described in the *Learning a Skill* activity.

Materials
- ✓ sentence strips
- ✓ dry elbow macaroni

LITERATURE LINKS

A Tree Is Nice by Janice May Udry

Dinosaurs to the Rescue by Laurie Krasny Brown and Marc Brown

Jack, the Seal, and the Sea by Gerald Aschenbrenner

Jack's Garden by Henry Cole

The Maple Tree by Millicent E. Selsam

Pablo's Tree by Pat Mora

Prince William by Gloria Rand

The Tree by Judy Hindley

Uncle Vova's Tree by Patricia Polacco

What Can We Do About Litter? by Donna Bailey

BUTTONS BUTTONS

Different qualities and attributes of buttons are explored through a photo essay and rhyming text.

BUTTONS ABOUND
Wall story

A bright idea and project from Kim Jordano and her kindergartners, Lee School, Los Alamitos, California

Fold a length of butcher paper in half horizontally, and then accordion-fold it to make pages. Help children write a button rhyme modeled after the pattern in the book. Sponge-paint T-shirt shapes and other designs, and attach them to the pages. Use construction paper scraps for art on the last page. Attach text and buttons and hang the wall story on the story clothesline for a great "point and read" activity. The wall story can also be cut apart and made into a class big book.

Materials
- ✓ butcher paper
- ✓ variety of buttons
- ✓ sponge-painting supplies
- ✓ construction paper scraps

Adapt this activity as a way for children to get hands-on practice with number combinations. Children can help make a number clothesline.

SCIENCE CONCEPTS

- Matter has many characteristics.

- Matter can be classified by characteristics such as color, shape, size, and quantity.

WRITING FRAMES

_____ buttons. _____ buttons.

_____ buttons. _____ buttons.

Buttons, buttons on my _____.

Buttons, buttons on my _____.

Buttons, buttons on my _____.

Buttons, buttons everywhere!

RELATED SKILLS

- parts of speech: *descriptive words (adjectives)*

- phonics: *initial consonant b, rhyming words (two/blue; big/pig; round/found)*

- sorting and classifying

MYSTERY BUTTON
Hands-on activity

Give each pair of children a bag of buttons. The level of difficulty for the activity is determined by how many buttons are in each bag and how similar the buttons are. Less buttons with unique characteristics are more appropriate for younger children.

Divide children into pairs. Partner A closes his or her eyes, and partner B chooses one button from the bag and places it in partner A's hand to feel and describe. Partner B then places the button back in the bag. Partner A opens his or her eyes and tries to identify the button among the others in the bag, using the sense of touch. Partners can take turns guessing different buttons. Pairs of players can swap bags.

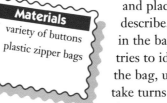

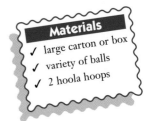

THE GREAT BALL SEARCH
Scavenger hunt

Give children a week to collect any balls they can find. Designate a box where they can place the balls each morning when they enter the room. At the end of the week, empty the box in the middle of the room and give children time to observe all the different kinds of balls, including their unique and common qualities.

Children can explore ways to group the balls according to characteristics such as size, color, types of games they're used for, and balls that bounce and balls that don't. Sort balls on the floor by placing them in hoola hoops. Extend this activity by crossing the hoops to make a Venn diagram.

LEARNING A SKILL

Descriptive words (adjectives)

Read the big book version of *Buttons Buttons* and point out the words used to describe the buttons. Explain that number, color, shape, or size words can be describing words. Have children brainstorm more describing words for buttons in the book and stick them on the pages with self-adhesive notes.

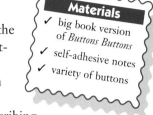

Materials
✓ big book version of *Buttons Buttons*
✓ self-adhesive notes
✓ variety of buttons

Play "Guess My Button." In pairs, children place several buttons on the tabletop. Player A tries to guess which button player B is thinking of by asking descriptive questions such as: *Is it shiny? Is it little? Does it have two holes?* The guesser can remove those sharing characteristics mentioned in questions that were answered, *No.*

LINKING SCHOOL TO HOME

Button bag

A bright idea from Kim Jordano, kindergarten teacher, Lee School, Los Alamitos, California

Make a button bag for children to take turns taking home. Sew colorful buttons to the outside of a cloth bag and label the bag using puffy paint or fabric markers. Place a copy of *Buttons Buttons* in the bag for the child to read to family members. Also include a plastic jar of buttons, an inexpensive plastic sectional tray, and a simple set of directions for using the bag at home.

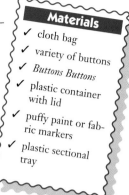

Materials
✓ cloth bag
✓ variety of buttons
✓ *Buttons Buttons*
✓ plastic container with lid
✓ puffy paint or fabric markers
✓ plastic sectional tray

LITERATURE LINKS

"A Lost Button" from *Frog and Toad Are Friends* by Arnold Lobel

The Button Box by Margaret Reid

Elephant Buttons by Noriko Ueno

I Spy (series) by Jean Marzollo

Large as Life by Julia Finzel

My Cat Likes to Hide in Boxes by Eve Sutton

Shapes by John Reiss

The Shapes Game by Paul Rogers

I CAN'T SLEEP

I can't sleep.
I feel something bumpy.

My marbles!

A child can't fall asleep because of the wild assortment of things he finds in his bed.

SCIENCE CONCEPTS

- Matter has observable characteristics.

- Matter can be described by texture, shape, and size.

- People use their senses to get information about matter.

- People use their senses to make predictions about matter.

- Matter takes up space.

WRITING FRAMES

I feel something _____.

My _____!

My _____ feels _____.

I can't sleep because _____.

I feel matter that is _____.

RELATED SKILLS

- real or make-believe

- picture details (clues)

- drawing conclusions

- return sweep

- sensory description

- punctuation: *exclamation points*

- contractions: *can't, isn't*

- parts of speech: *nouns, descriptive words (adjectives)*

A*CTIVITY* — BRING IN YOUR BEAR!
Bar graph

A bright idea from Rosa Drew, first grade teacher, Vista Verde School, Irvine, California

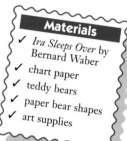

Materials
✓ *Ira Sleeps Over* by Bernard Waber
✓ chart paper
✓ teddy bears
✓ paper bear shapes
✓ art supplies

Read *Ira Sleeps Over* to the class. Compare and contrast the book with *I Can't Sleep*. Have each child bring in a teddy bear. (Provide extras for children to borrow, if needed.) Have children gather on the floor with their bears. Line the bears up by color, size, and other characteristics sparked by discussion (clothes/no clothes, fuzzy/not fuzzy). Discuss your findings.

There are more brown bears than white bears.

Invite children to color paper bear shapes to represent their own bears. Each child can glue his or her bear on the graph. When the graph is complete, discuss the results.

Our Bears Are Just Right!

pink	🐻				
white	🐻	🐻			
brown	🐻	🐻	🐻	🐻	🐻
black	🐻	🐻	🐻		

A WE FEEL MATTER
Interactive display

A bright idea and project from Susan Gardner and her K–1 students, Rolling Hills School, Fullerton, California

Materials
- ✓ 8 sheets of 12" x 18" tagboard
- ✓ wide cloth tape
- ✓ chart
- ✓ various objects of different textures
- ✓ art supplies

Tape the tagboard together lengthwise. Add tape to both sides and accordion-fold the display so it stands up on a counter or table. Brainstorm with children a list of words to describe how things feel and write them as headers on a chart. Underneath each header, have children list common objects that fit the header description.

Transfer the text from the chart onto the sections of the tagboard display. Glue objects that fit each category on the display along with children's illustrations. In front of the display, place objects for children to experience, using their sense of touch.

A WHAT'S UNDER MY BED?
Writing and art project

A bright idea from Rosa Drew, first grade teacher, Vista Verde School, Irvine, California

Have children follow these directions:

1. Fold a sheet of tagboard in half.

2. On the top half, make a bed from construction paper scraps.

3. Attach a rectangular piece of fabric to the bed so it can be lifted.

4. Draw something hidden under the bed and add people and other details to your picture.

5. Write or dictate a short paragraph describing what's under the bed so the reader can guess. Attach this paper to the bottom half of the project.

Materials
- ✓ fabric scraps
- ✓ construction paper scraps
- ✓ art supplies
- ✓ 12" x 18" tagboard
- ✓ writing paper

What's under my bed?

MY CAT

Something is under my bed. It is making scratching noises. It feels furry. It sounds like a motor. What is it?

LEARNING A SKILL

Real or make-believe
Picture details

Read the big book version of *I Can't Sleep* while children pay special attention to the pictures. Notice the little stories within the story about the spaceship and frog.

Materials
- ✓ big book version of *I Can't Sleep*
- ✓ pocket chart
- ✓ yarn
- ✓ sentence strips
- ✓ stickers
- ✓ art supplies

Divide a pocket chart in two sections using a piece of yarn. Label the sections *Real* and *Make-believe*. Have children write or dictate sentences onto sentence strips describing details in the pictures. Invite them to place the sentences under the appropriate section of the pocket chart.

Make this a self-checking activity by placing a sticker on the back of each sentence strip that matches the stickers you place on the labels *Real* or *Make-believe*.

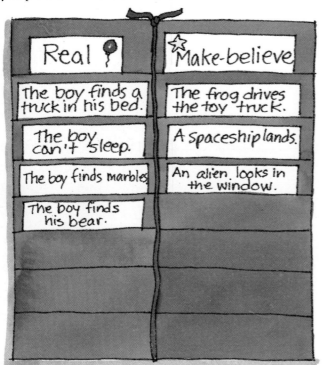

LINKING SCHOOL TO HOME

Traveling bag

Prepare a bed-shaped book with one page per child, including the text: "When I can't sleep I _____." Send this book home each day with a different child. Place it in a bag or small suitcase with a copy of *I Can't Sleep* and supplies such as fabric scraps, glitter glue, crayons, and wrapping paper. Together, families can read the book, color and decorate the bedspread, and finish the writing frame. They can also read and enjoy the other pages completed by classmates. Invite each child to share his or her family's work with the class the next day.

Materials
- ✓ drawing paper
- ✓ tote bag or small suitcase
- ✓ art supplies
- ✓ collage materials

LITERATURE LINKS

All Shapes and Sizes by Shirley Hughes

Aunt Nina, Good Night by Franz Brandenberg

Goodnight, Goodnight by Eve Rice

Goodnight Moon by Margaret Wise Brown

Half a Moon and One Whole Star by Crescent Dragonwagon

Ira Sleeps Over by Bernard Waber

"Strange Bumps" from *Owl at Home* by Arnold Lobel

Ten, Nine, Eight by Molly Bang

You Be Good and I'll Be Night by Eve Merriam

I SEE COLORS

A variety of common objects are grouped together by color and function.

COLOR MUSEUM

Hands-on sorting activity

A bright idea from Marlene Beierle and Anne Sylvan, multi-age teachers, Olde Orchard School, Columbus, Ohio

Materials

✓ tagboard or table-top divided into eight sections
✓ color word cards
✓ shoe box or laundry basket

Invite children to gather small objects from home or the classroom that represent colors on the color cards. Place all the objects in a shoe box or laundry basket. Have children place a color card on each section of the table or tagboard and take turns placing objects in the correct sections.

As an extension, you may wish to time the activity. See how many objects children can place correctly in one minute. Then invite children to study the items in one section for one minute, take them away, and see how many they can recall.

SCIENCE CONCEPTS

- Matter is anything that has weight and takes up space.

- Matter can be classified by color, shape, and function.

- People can use their senses to observe and classify matter.

WRITING FRAMES

I see _____.

A _____ is (color word).

I see a (color word) (name of object).

I see a _____ as _____ as the _____.

RELATED SKILLS

- sorting and classifying

- vocabulary development

- high-frequency words: *I, see*

- parts of speech: *descriptive words (adjectives)*

- phonics: *ee (see), initial consonant s*

COLOR EXPLORATION
Experiment with color

Place the materials listed in a center. One or two children at a time can explore mixing various color combinations.

COLOR DAYS
Class book

Celebrate colors with "Color Days." Have children dress in clothes for the color of the day and bring in one item of that color from home. Take photographs of children with the items. Use the photos to make a crayon-shaped book. A colorful snack will add to the fun!

Save the items from each color day. As a culminating activity, mix all the items together. Divide them among small groups. Children can sort the items by characteristics such as color, size, shape, and use.

COLOR CHEMISTRY
Science experiment

Tear red cabbage leaves into small pieces. Bring a pan of water to a boil, remove from heat, and place the cabbage in the pan. After soaking for half an hour, pour the violet-colored water into a pitcher, and discard the cabbage. Assign children to groups of three or four, and have each group con-

duct the following investigation:

Fill four clear plastic cups half full with the cabbage water, plain water, white vinegar, and water mixed with baking soda. Pour a little of the cabbage water into the other cups. Record the results on an observation sheet.

When all groups have finished, gather together and discuss the results. The violet cabbage water turns red in acid (vinegar), green in alkaline (baking soda), and stays the same in neutral water.

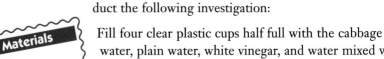

LEARNING A SKILL

Sorting and classifying
Vocabulary development

A bright idea and project from
Linda Goodner and her kinder-
gartners, Lee School, Los Alamitos,
California

Read the big book version of *I
See Colors* and discuss how objects
are grouped by color (e.g., red
toys, blue kitchen items). Help children brainstorm other
ways objects can be grouped. You might start with ways
things are grouped in the classroom, such as art supplies,
math manipulatives, and writing materials.

Choose one category to explore for a class book entitled
We See Colors. For example, choose a clothing category and
list different clothing on the board. Assign a color to each
kind of clothing and use this information with the writing
frame "We see (color word) (kind of clothing)" to create
text for the book. Have children illustrate each page.

LINKING SCHOOL TO HOME

Scavenger hunt

Send home the book *I See Colors* or
the class book *We See Colors* (from
the *Learning a Skill* activity) with
one child each day. In the tote
bag, include a timer and a short
set of directions to send family members on a
five-minute scavenger hunt. Have them look for certain
colored objects. When the timer goes off, family members gather and count each other's finds. They can also
sort the objects into groups such as red toys, red kitchen
items, or red clothing.

LITERATURE LINKS

Afro-Bets Book of Colors by Margery W. Brown

All the Colors of the Race by Arnold Adoff

Color by Ruth Heller

The Color Box by Dayle Ann Dodds

Colors Everywhere by Tana Hoban

The Crayola® Counting Book by Rozanne Lanczak Williams

Greens by Arnold Adoff

Growing Colors by Bruce McMillan

Is It Red? Is It Yellow? Is It Blue? by Tana Hoban

Mouse Paint by Ellen Stoll Walsh

Of Colors and Things by Tana Hoban

Samuel Todd's Book of Great Colors by E.L. Konigsburg

Seven Blind Mice by Ed Young

Who Said Red? by Mary Serfozo

WHAT'S IN MY POCKET?

The fun things inside children's pockets are described through a simple, repetitive chant.

SCIENCE CONCEPTS

• Matter can be classified according to different properties.

• Matter can be described by texture, shape, and size.

• People use their senses to get information about matter.

WRITING FRAMES

Pocket, pocket, what's in my pocket?

Something that's _____.

It's a _____.

A _____ is _____ and _____.

RELATED SKILLS

• parts of speech: *descriptive words (adjectives)*

• contractions: *what's, that's*

• punctuation: *question marks, commas*

• drawing conclusions

• return sweep

ARE THEY THE SAME?
Comparing matter

Pour the same amount of oil and water respectively in containers A and B. (Tint the water slightly with yellow food coloring so it looks like the oil.)

Give children time to investigate the two containers and guess whether or not they contain the same liquid. Ask them to support their answers.

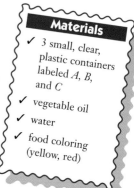

Materials
✓ 3 small, clear, plastic containers labeled *A, B,* and *C*
✓ vegetable oil
✓ water
✓ food coloring (yellow, red)

Add red food coloring to the water in container B. Ask children if anything except the color of the liquid in container B has changed. Then pour half of the liquid in containers A and B into container C, and have children note the results. Let container C sit undisturbed for a couple of minutes and observe the results again.

Discuss what conclusions can be drawn from the experiment. Were the contents of containers A and B the same? Why did the color of container B's contents change when red food coloring was added? Which liquid weighed the least? How do you know?

Encourage children to write and draw about the experiment in their science journals.

THE MYSTERY CAN
Guessing game

Send home the
"mystery can"
with a different child
each day. Invite the child to find
an object at home that fits in the
can and bring it to school the
next day. Have him or her give clues
to the class until someone guesses what's inside.

Materials
✓ metal tin with lid

POCKET APRON GAME
Whole-group activity

Place small objects in a box as children
watch. Have one child put on the apron and
place one of the objects into the apron pocket
without the other children seeing. He
or she can chant: *Pocket, pocket!
What's in my pocket? Something
that's _____.*

Have children take turns guessing
what's in the pocket. The chant
can be repeated with other clues
until someone guesses correctly.
That child wears the apron next.

Materials
✓ apron with pocket
✓ variety of small objects
✓ box

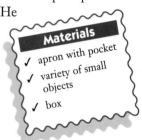

WHAT'S THE MATTER?
Hands-on activity

Wrap the soap, sponge, and block separately in wrapping paper. Invite children to handle each package and try to guess what's inside. Remind them that different matter can sometimes share some of the same characteristics. List what is the same about each package and what is different. When children think they know what is inside each package, discuss how they determined their guesses. Discuss how we use our five senses to describe matter.

Materials
✓ bar of soap, sponge, block of wood (similar in size)
✓ wrapping paper

LEARNING A SKILL

Descriptive words (adjectives)

A bright idea and project from Tebra Corcoran and her kindergartners, Lee School, Los Alamitos, California

Ask each child to create a page for a class book by decorating a pocket pattern or the back of an envelope and gluing it to construction paper "blue jeans." Children should be careful to apply glue only to outside edges of the "pocket," excluding the top so that a small object can be placed inside.

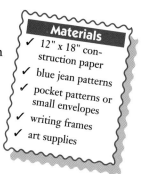

Materials
✓ 12" x 18" construction paper
✓ blue jean patterns
✓ pocket patterns or small envelopes
✓ writing frames
✓ art supplies

Have children glue the jeans to a page with the writing frame. Invite children to finish the frame after putting a small object in the pocket. Explain that the clues they write about their object are describing words and will help the reader guess the object. Combine all the pages into a guessing book.

LINKING SCHOOL TO HOME

Take-home backpack

A bright idea from Trisha Callella, kindergarten teacher, Rossmoor School, Los Alamitos, California

Materials
✓ jeans
✓ variety of objects (marble, block, feather, rock, small teddy bear, plastic worm)

Make a backpack out of jeans with several pockets. Send the backpack home with a different child each day. In the backpack place a copy of *What's In My Pocket?*, a marble, block, feather, rock, teddy bear, worm, and a set of directions. As the child reads the story, he or she can give a clue and ask the parent to look in the pockets of the backpack for the object described. They continue through the book, then reverse roles.

stitch up legs

LITERATURE LINKS

Allsorts by Tony Wells

The Biggest, Smallest, Fastest, Tallest by Robert Lopshire

Circles, Triangles, and Squares by Tana Hoban

Color Zoo by Lois Ehlert

Grandfather Tang's Story by Ann Tompert

Is It Red? Is It Yellow? Is It Blue? by Tana Hoban

Many Luscious Lollipops: A Book About Adjectives by Ruth Heller

Ohmygosh My Pocket by Janet Perry Marshall

Shapes: How Do You Say It? by Meredith Dunham

Shapes, Shapes, Shapes by Tana Hoban

IT'S MELTING!

A book about
things that melt.

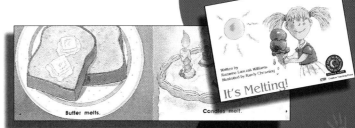

WHAT MELTS IN THE SUN?

Interactive mural

A bright idea and project from Cori Giacchino and her summer school kindergartners, Hopkinson School, Los Alamitos, California

Materials
✓ butcher paper
✓ tempera paint
✓ paintbrushes
✓ construction paper

Help children gather various items and predict whether they will melt in the sun. Chart children's predictions before placing the items outside in the sun for a specified amount of time. Record the results by making a mural. Children can paint and cut out different objects from the experiment and attach them to butcher paper. Add the title *What Will Melt in the Sun?*, and have children add folded flaps cut from construction paper next to each item on the mural. On the top, write the name of the object. When the flap is lifted the answer "Yes" or "No" appears. Paintings of objects that did not fit on the mural can be used to illustrate a class book.

SCIENCE CONCEPTS

• Properties of matter can change.

• Changes in matter can be observed.

• Matter reacts to changes in the environment.

• Matter reacts to energy input.

WRITING FRAMES

_____ melts.

_____ is melting in the _____.

RELATED SKILLS

• cause and effect

• phonics: *initial consonant m*

• parts of speech: *nouns, verbs (melt/melts)*

THE GREAT ICE CUBE RACE
Melting experiment

Have children investigate what causes melting and what would affect the rate of melting. Have pairs of children place two ice cubes in separate bags and decide on two different places to put them. Set the timer and have children check their ice cubes every five minutes and record changes. At the end of the activity, chart everyone's observations and discuss the results.

Materials
- ✓ ice cubes (two per pair)
- ✓ plastic zipper bags
- ✓ timer

WHAT MELTS THE FASTEST?
Making predictions

Invite children to guess which will melt first—an ice cube, a lit birthday candle, or a scoop of ice cream. Record children's guesses and then display the materials and let them melt. Extra care should be taken with the lighted birthday candle (teacher handling only). Children can draw and write about the results.

Materials
- ✓ ice cube
- ✓ birthday candle
- ✓ matches
- ✓ scoop of ice cream

MELTDOWN!
Art project

A bright idea from Judi Hechtman, first grade teacher, University School, Indiana, Pennsylvania

Show children how to carefully scrape various colored crayons onto squares of waxed paper using open scissors. Have them place another sheet of waxed paper over the crayon scrapings and press with a warm iron. Children should be closely supervised by an adult when using the iron. The crayon scrapings will melt and blend together. Make construction paper frames and hang the designs in the classroom windows.

Materials
- ✓ crayons (paper removed)
- ✓ waxed paper
- ✓ construction paper
- ✓ scissors
- ✓ iron

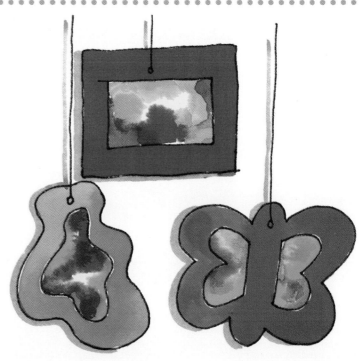

IT'S FREEZING!
Hands-on activity

Give pairs of children two containers and ask them to fill both with one of the liquids. Ask children to label each container with its contents and partners' names. Freeze one container from each pair overnight, leaving the other at room temperature as the "control" sample. The next day, assign children into groups of four (two pairs) so they can share results. By freezing only one liquid, the children have the means to make an accurate comparison. Observe, chart, and discuss changes that did or did not occur.

Materials
- ✓ small plastic containers with lids
- ✓ various liquids (water, soda, vinegar, shampoo, rubbing alcohol, peroxide)

LEARNING A SKILL

Cause and effect

Read the big book version of *It's Melting!* and discuss what is causing items to melt in each picture. Have children dictate sentences, giving causes for each "meltdown" as you write them on chart paper or sentence strips. Reread the responses together and have children come up with new situations not pictured in the book. Use the text for a class-illustrated big book or pocket chart set.

Materials
- ✓ big book version of *It's Melting!*
- ✓ chart paper or sentence strips
- ✓ pocket chart
- ✓ art supplies

LINKING SCHOOL TO HOME

Dancing raisins

Send home the following letter to parents:

Dear Parents,

We've been studying the three states of matter (solid, liquid, and gas). Please have your child share how much he or she has learned and do this simple activity together.

Dancing Raisins

1. *Pour any clear soda, such as ginger ale (LIQUID), into a tall clear glass.*

2. *Drop in ten raisins. (SOLID)*

3. *The bubbles in the soda (GAS) will make the raisins appear to "dance" as they rise to the top and then drop again.*

Materials
- ✓ take-home letter
- ✓ clear soda
- ✓ raisins
- ✓ tall clear glass

LITERATURE LINKS

Air by David Bennett

The Air Around Us by Elenore Schmid

Benny Bakes a Cake by Eve Rice

I Spy (series) by Jean Marzollo

The Important Book by Margaret Wise Brown

My First Science Book by Angela Wilkes

The Sun by Michael George

Water's Way by Lisa Westberg Peters

MOM CAN FIX ANYTHING

Mom helps a little girl fix a variety of broken objects.

ACTIVITY

TOOLS MAKE OUR LIVES EASIER
Wall story

A bright idea and project from Mary Kay Stephens and her second grade summer school class, Rossmoor School, Los Alamitos, California

Materials
- ✓ *Tools* by Ann Morris
- ✓ butcher paper
- ✓ construction paper
- ✓ art supplies

Read *Tools* and other books about tools. Brainstorm a list of ways tools help us in our everyday lives. Set up a learning center where children can explore several tools. Using the frame "Tools help us _____," have children write and illustrate a wall story about tools.

ACTIVITY

WE CAN FIX ANYTHING!
Working as a team

Make a list of things in the classroom that need to be fixed or jobs that need to be done. Pairs of children can choose jobs and report to the class on how the job was accomplished and what tools were used.

Materials
- ✓ chart paper
- ✓ variety of tools

WHAT TOOL WILL YOU USE?
Hands-on activity

Give each child a paper cup and a handful of cereal. Ask children to place the cereal in the cup without using their hands. They may choose from the tools available or invent their own. Children will conclude that different tools can do the same job and that the way an object is used determines whether or not it is a tool.

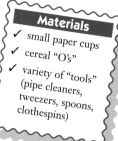

Materials
- ✓ small paper cups
- ✓ cereal "O's"
- ✓ variety of "tools" (pipe cleaners, tweezers, spoons, clothespins)

SWAPPING TOOLS
Tools and their uses

Have each child fold a sheet of drawing paper to form four or eight squares. In each square on the left side, children can draw and color a different tool. Have children swap papers with a friend and draw ways the tools are used in the empty squares on the right.

Materials
- ✓ 12" x 18" drawing paper
- ✓ crayons or markers

LET'S BUILD IT!
Classroom carpentry center

Set up a classroom carpentry center for hands-on learning about tools and building. Explain the use of tools and review safety precautions. Include directions for simple projects. Display children's finished projects in the classroom.

Materials
- ✓ wood scraps (pine, balsa)
- ✓ tools (hammer, nails, glue, sandpaper, measuring tape, screwdriver)
- ✓ collage materials

LEARNING A SKILL

Vocabulary development
Creative thinking

Divide children into small
groups of two to four, and give
each group one gadget. Ask
them to work together to figure
out what the tool might be
used for. Encourage them to
brainstorm multiple ideas to report to the
class. Afterwards, introduce the tool name and its most
common use. Children usually come up with much more
creative uses for the tools!

Materials
✓ several uncommon household tools or gadgets (garlic press, nutcracker, melon baller)

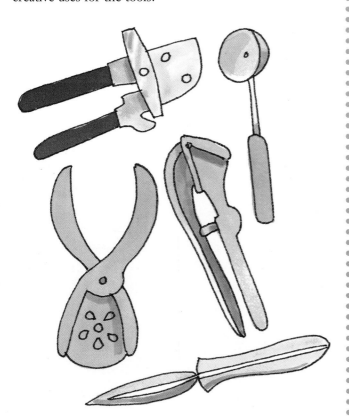

LINKING SCHOOL TO HOME

Home improvement

Share the class study of tools
with parents. Ask parents to
include their children in the
process of fixing a broken object
in their home. Remind them to
discuss the tools needed, the necessary steps, and any vocabulary
unique to the process. Ask children to draw a picture illustrating how they completed the task. Have children give
an oral report explaining how they helped.

Materials
✓ broken object
✓ tools
✓ crayons or markers

LITERATURE LINKS

Bathtubs, Slides, Roller Coasters: Simple Machines That Are Really Inclined Planes by Christopher Lampton

Hazel's Amazing Mother by Rosemary Wells

Just Us Women by Jeanette Caines

Marbles, Roller Skates, Doorknobs: Simple Machines That Are Really Wheels by Christopher Lampton

My First Book of How Things Are Made by Greg Jones

Three Brave Women by C.L.G. Martin

The Toolbox by Anne Rockwell

Tools by Ann Morris

The Visual Dictionary of Everyday Things published by Dorling Kindersley

Whirr Pop Click Clang by Carol MacKenzie

ON THE GO

This book showcases various
forms of transportation.

ON THE GO!
Class pop-up book

A bright idea and project from Eileen Young and her first graders,
Lincoln School, Long Beach, California

Follow the directions for *Read the Clues* on page 19 to make a class book about transportation. Have each child create one page for the book using the writing frame "A _____ takes me to the _____."

Materials
✓ drawing paper
✓ construction paper
✓ art supplies
✓ 12" x 18" heavy
  paper (for cover)

SCIENCE CONCEPTS

• People move themselves and things from place to place using various forms of transportation.

• Walking is a form of transportation.

• People have invented different means of transportation to travel over land, air, and water.

WRITING FRAMES

A _____ takes me to the _____.

I can ride _____.

I get to the _____ in a _____.

I went to _____ in/on a _____.

RELATED SKILLS

• high-frequency words: *a, me, to, the*

• vocabulary development

• parts of speech: *nouns, verbs (take/takes)*

• possessive: *Grandpa's*

HOW DO WE TRAVEL?
Transportation mural and chart

Make a class mural illustrating modes of transportation in the air, on land, and on water. Add labels to the mural or write a separate chart describing the pictured modes of transportation. Children can refer to the mural when writing about transportation.

Materials
- ✓ butcher paper
- ✓ construction paper
- ✓ art supplies
- ✓ collage materials

Mural by Eileen Young's first graders.

TRANSPORTATION INVENTIONS
Hands-on project

Using various materials, have children create their own form of transportation. They can test their inventions in the block center or at the water table and dictate or write sentences about them. Some children may make a simple car, while others make a floating, flying machine! In small groups, invite children to share their creations, describing how they made them and how they work.

Materials
- ✓ "building" materials (toothpicks, walnut shells, Styrofoam trays, milk cartons, wood scraps, bottle tops, corks)
- ✓ art supplies

TRANSPORTATION EXPLORATION
Free exploration

Place the listed materials and other related items at a learning center. Invite children to explore through free play. Then ask them to draw one item and write about how it works and how it was made for travel. Hang completed papers at the center.

Materials
- ✓ toy vehicles (car, truck, plane, bus, boat)
- ✓ pan of water
- ✓ roller skate
- ✓ tennis shoe
- ✓ blocks
- ✓ drawing paper
- ✓ art supplies

HOW DOES THE MAIL GET WHERE IT'S GOING?

Class research

A bright idea and project from Eileen Young and her first graders, Lincoln School, Long Beach, California

Materials
- ✓ chart paper
- ✓ art supplies

Plan a class field trip to the post office. The focus of the trip will be to find out what modes of transportation are used to deliver mail. Prepare beforehand and brainstorm questions to ask the tour guide. Assign volunteers to take notes. After returning to the classroom, create a story map showing how mail is delivered. Use the story map as a resource when creating a class book.

LEARNING A SKILL

High-frequency words

A bright idea and project from Eileen Young and her first graders, Lincoln School, Long Beach, California

Materials
- ✓ blackline masters
- ✓ construction paper
- ✓ library pockets
- ✓ art supplies

Supply children with copies of blackline masters shown in the photograph. Have children finish the art and fill in the blanks with the high-frequency words *a, me, to, the,* and the verb *takes.* Children can make a simple cover from construction paper and make a paper doll to use when reading the book. Store the dolls in library pockets glued to the inside front cover. Children can practice one-to-one matching by using the paper doll to point to each word as they read.

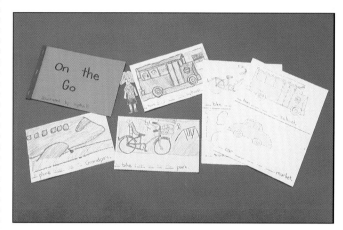

LINKING SCHOOL TO HOME

How'd you get there?

Materials
- ✓ take-home form

Send home a form with children asking parents to help list all types of transportation used by the family in the past year. Request that children bring their forms back to share or to put in a class book.

LITERATURE LINKS

Airplanes; Airport; Boats; I Want to Be an Astronaut; Trains; Trucks by Byron Barton

Boats; Cars; Planes; Trucks; Things That Go by Anne Rockwell

Flying; Freight Train; School Bus; Truck by Donald Crews

Jessie's Journey by Angela McAllister

On the Go by Ann Morris

The Post Office Book: Mail and How It Moves by Gail Gibbons

This Is the Way We Go to School by Edith Baer

The Train by David McPhail

Trains by Gail Gibbons

Truck Song by Diane Siebert

WHAT HAPPENED?

Water is shown in three
forms—solid, liquid, and
gas—throughout the seasons.

SCIENCE CONCEPTS

- Matter exists in three states—solid, liquid, and gas.

- The addition or removal of heat affects the states of matter.

- Some solids can be dissolved in liquid.

WRITING FRAMES

What happened to the
_____?

It _____.

RELATED SKILLS

- drawing conclusions

- cause and effect

- punctuation: *question marks*

- high-frequency words: *what, to, the, and, then, I*

- question and answer format

ACTIVITY

WHY DID IT HAPPEN?
Cause and effect

A bright idea and project from Katy Dunbar
and her first graders, Portola Hills School,
Trabuco Canyon, California

Divide a class chart into two columns and add the headings
What Happened? and *Why?* Reread the big book version of
What Happened? to the class and discuss changes that
occurred throughout the book. List changes on the left
column of the chart. Invite children to complete the right side of the
chart with reasons *why* the changes occurred. Use the chart to create an innovation
to the book. Children can illustrate the book using cut and torn paper.

Materials

✓ 12" x 18" white construction paper

✓ colored construction paper scraps

✓ big book version of *What Happened?*

What Happened?	Why?
The water froze. The ice melted. The water fogged up the window.	The air got cold. The sun came out. It condensed.

SOLID, LIQUID, GAS
Science experiment

Divide children into small groups of two or three. They will combine a solid and liquid to form gas. Prepare bottles ahead of time by filling them with vinegar. Distribute the rest of the materials. Have each group use a funnel to put 2 tablespoons of baking soda into a balloon. They then twist the balloon and put the end over the bottle. When children untwist the balloon, the baking soda drops into the vinegar and gas will be produced, inflating the balloon.

Materials
- ✓ plastic bottles
- ✓ balloons
- ✓ baking soda
- ✓ vinegar
- ✓ funnels

SOLID TO LIQUID
Yummy activity

During your study unit on matter, plan daily snacks that demonstrate solids melting to liquid. Here are some suggested treats:

Materials
- ✓ ice
- ✓ bouillon
- ✓ hot water
- ✓ ice cream
- ✓ milk

- room-temperature juice with ice
- bouillon or other instant soups in hot water
- ice cream mixed with milk to make milkshakes

WHAT DISSOLVES IN WATER?
Learning center activity

Fill cups with water. Stir a small amount of each material into different cups and label the cups. Record results on an observation sheet.

Materials
- ✓ clear plastic cups
- ✓ water
- ✓ materials to dissolve (salt, fine sand, tea leaves, baking soda, soil, sugar, instant drink powder)

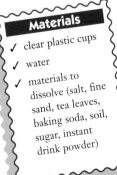

LEARNING A SKILL

Drawing conclusions
Art project

A bright idea and project from Judi Hechtman, Marjorie Mambo, and their first graders, University School, Indiana, Pennsylvania

> **Materials**
> ✓ salt
> ✓ tempera paint
> ✓ paintbrushes
> ✓ drawing paper
> ✓ clear plastic containers

Help children dissolve salt in clear plastic containers filled with liquid tempera paint. They might notice (depending on the thickness of the paint) that there is a limit as to how much salt will dissolve before the excess starts falling to the bottom. (This is called the saturation point.) Invite children to paint designs with the "salt paint" and observe what happens as the paint dries. Discuss the results and display the beautiful salt crystal paintings.

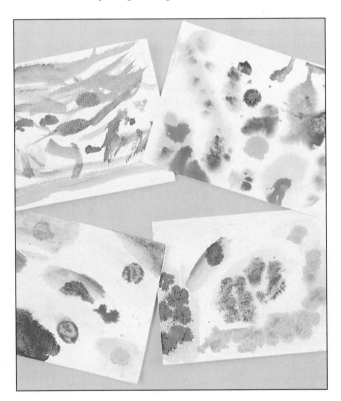

LINKING SCHOOL TO HOME

Mystery matter

Ask parents to help their children go on a scavenger hunt at home and make a list of five solids, five liquids, and five gases they find. Supply parents with the following recipe for Glop which isn't really a solid or a liquid—it's just plain weird!

Glop

1 cup cornstarch
1/2 cup water
food coloring (optional)
plastic zipper bag

1. Pour water into a bowl and add cornstarch little by little while stirring. Add a few drops of food coloring to make colored Glop.

2. Keep mixing until the Glop is well blended and store in a plastic zipper bag.

LITERATURE LINKS

Discovering Science Secrets by Sandra Markle

Easy Science Experiments by Diane Molleson and Sarah Savage

Experiments with Water by Bryan Murphy

Heat, Light, and Sound: Energy at Work by Melvin Berger

175 Science Experiments to Amuse and Amaze Your Friends by Brenda Walpole

Simple Science Experiments with Everyday Materials by Muriel Mandell